NASA NEWS

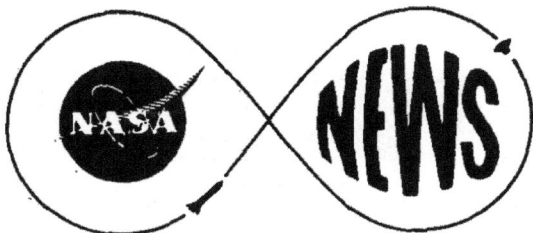

NATIONAL AERONAUTICS AND SPACE ADMINISTRATION
Washington, D. C. 20546
202-755-8370

RELEASE NO: 72-64K

FOR RELEASE:
THURSDAY A.M.
April 6, 1972

PROJECT: APOLLO 16
(To be launched no
earlier than April 16)

APOLLO 16
YOUNG · MATTINGLY · DUKE

P R E S S
K I T

contents

- more -   March 22, 1972

- more -

TABLES AND ILLUSTRATIONS

- more -

Cover: Apollo 16 Landing
This section of a panoramic photograph consisting of 27 separate frames taken by
Charles Duke shows the Apollo 16 landing site in the lunar highlands, April 23, 1972.

Image Credit: NASA

Published by Books Express Publishing
Copyright © Books Express, 2012
ISBN 978-1-78039-865-5

Books Express publications are available from all good retail and online booksellers. For
publishing proposals and direct ordering please contact us at: info@books-express.com

NATIONAL AERONAUTICS AND
SPACE ADMINISTRATION
Washington, D. C. 20546
Phone: (202) 755-8370

Kenneth C. Atchison
(Phone: 202/755-3114)

FOR RELEASE:
THURSDAY A.M.
April 6, 1972

RELEASE NO: 72-64

APOLLO 16 LAUNCH APRIL 16

Apollo 16 (scheduled for an April 16 launch) will
devote its 12-day duration to gathering additional knowledge about the environment on and around the Moon and about
our own planet Earth.

During the three days two Apollo 16 crewmen spend on the
lunar surface north of the crater Descartes, they will
extend the exploration begun by Apollo 11 in the summer
of 1969 and continued through the Apollo 12, 14, and 15
lunar landing missions. In addition to gathering samples
of lunar surface material for analysis on Earth, the crew
will emplace a fourth automatic scientific station.

- more -

March 22, 1972

An extensive array of scientific experiments in the orbiting command/service module will search out and record data on the physical properties of the Moon and near-lunar space and photographic images to further refine mapping technology. Additionally, the command module pilot will photograph astronomical phenomena in the distant reaches of space.

The Descartes landing site is a grooved, hilly region which appears to have undergone some modification by volcanic processes during formation. The Descartes region is in the southeast quadrant of the visible face of the Moon and will offer an opportunity to examine several young, bright-rayed craters created by impacts in the volcanic terrain.

John W. Young is Apollo 16 mission commander, with Thomas K. Mattingly flying as command module pilot and Charles M. Duke, Jr. as lunar module pilot. Young is a US Navy captain, Mattingly a Navy lieutenant commander, and Duke a US Air Force lieutenant colonel.

- more -

Young and Duke will climb down from the lunar module onto the lunar surface for three seven-hour periods of exploration and experimentation. A major part of the first EVA will be devoted to establishing the nuclear powered, automatic scientific station -- Apollo Lunar Surface Experiment Package (ALSEP) -- which will return scientific data to Earth for many months for correlation with data still being returned by the Apollo 12, 14 and 15 ALSEPs.

The second and third EVAs will be devoted primarily to geological exploration and sample gathering in selected areas in the vicinity of the landing site. As in past missions, the crew's observations and comments will be supplemented by panoramic, stereo, and motion picture photographic coverage and also by television coverage. Crew mobility again will be aided by the use of the lunar roving vehicle.

In lunar orbit, Mattingly will operate experiments in the scientific instrument module (SIM) bay for measuring such things as the lunar surface chemical composition, and the composition of the lunar atmosphere. A high-resolution camera and a mapping camera in the SIM bay will add to the imagery and photogrammetry gathered by similar cameras flown on Apollo 15. Mattingly will perform an inflight EVA during transearth coast to retrieve film cassettes from these cameras.

- more -

Using hand-held cameras, Mattingly will photograph such phenomena in deep space as the Gegenschein, and looking earthward, photograph the ultraviolet spectra around Earth.

A second subsatellite, similar to the one flown on Apollo 15, will be ejected into lunar orbit to measure the effect of the Earth's magnetosphere upon the Moon and to investigate the solar wind and the lunar gravity field.

Apollo 16 is scheduled for launch at 12:54 pm EST April 16 from the NASA Kennedy Space Center's Launch Complex 39, with lunar landing taking place on April 20. The landing crew will remain at Descartes for 73 hours before returning to lunar orbit and for rendezvous with the orbiting command module on April 23. Earth splashdown will occur on April 28 at 3:30 pm EST at 5 degrees north latitude and 158.7 degrees west longitude in the central Pacific just north of Christmas Island. The prime recovery vessel, USS Ticonderoga, an aircraft carrier, will be located near the splashdown point to recover the crew and spacecraft.

Communications call signs to be used during Apollo 16 are "Casper" for the command module and "Orion" for the lunar module. The United States flag will be erected on the lunar surface in the vicinity of the lunar module, and a stainless steel plaque engraved with the landing date and crew signatures will be affixed to the LM front landing gear.

Apollo 16 backup crewmen are civilian Fred W. Haise, Jr., commander; USAF LtCol Stuart A. Roosa, command module pilot; and USN Captain Edgar D. Mitchell, lunar module pilot.

- more -

COUNTDOWN

The Apollo 16 launch countdown will be conducted by a government-industry team working in two control centers at the Kennedy Space Center (KSC).

Overall space vehicle operations will be controlled from Firing Room No.1 in the Complex 39 Launch Control Center. The spacecraft countdown will be run from an Acceptance Checkout Equipment control room in the Manned Spacecraft Operations (MSO) Building.

Extensive checkout of the launch vehicle and spacecraft components are completed before the space vehicle is ready for the final countdown. The prime and backup crews participate in many of these tests, including mission simulations, altitude runs, a flight readiness test and a countdown demonstration test.

The Apollo 16 rollout -- the 5.5-kilometer (3.4-nautical-mile) trip from the Vehicle Assembly Building (VAB) to the launch pad -- took place Dec. 13, 1971. Due to a problem associated with the fuel tank system in the command module reaction control system, the space vehicle was returned to the VAB on Jan. 27. The spacecraft was taken to the MSO Building for changeout of the tanks. After re-mating the spacecraft with the launch vehicle, the KSC team again rolled Apollo 16 back to Pad A on Feb. 9, 1972.

Apollo 16 will be the tenth Saturn V launched from Pad A (eight manned). Apollo 10 was the only launch from Pad B, which will be used again in 1973 for the Skylab program.

The Apollo 16 precount activities will start at T-6 days. The early tasks include electrical connections and pyrotechnic installation in the space vehicle. Mechanical build-up of the spacecraft is completed, followed by servicing of the various gases and cryogenic propellants (liquid oxygen and liquid hydrogen) to the CSM and LM. Once this is accomplished, the fuel cells are activated.

The final countdown begins at T-28 hours when the flight batteries are installed in the three stages and instrument unit of the launch vehicle.

At the T-9 hour mark, a built-in hold of nine hours and 54 minutes is planned to meet contingencies and provide a rest period for the launch crew. A one hour built-in hold is scheduled at T-3 hours 30 minutes.

Following are some of the highlights of the latter part of the count:

T-10 hours, 15 minutes	Start mobile service structure move to park site.
T-9 hours	Built-in hold for nine hours and 54 minutes. At end of hold, pad is cleared for LV propellant loading.
T-8 hours, 05 minutes	Launch vehiclé propellant loading – Three stages (LOX in first stage, LOX and LH_2 in second and third stages). Continues thru T-3 hours 38 minutes.
T-4 hours, 15 minutes	Flight crew alerted.
T-4 hours, 00 minutes	Crew medical examination.
T-3 hours, 30 minutes	Crew breakfast.
T-3 hours, 30 minutes	One-hour built-in hold.
T-3 hours, 06 minutes	Crew departs Manned Spacecraft Operations Building for LC-39 via transfer van.
T-2 hours, 48 minutes	Crew arrival at LC-39.
T-2 hours, 40 minutes	Start flight crew ingress.
T-1 hour , 51 minutes	Space Vehicle Emergency Detection System test (Young participates along with launch team).
T-43 minutes	Retract Apollo access arm to standby position (12 degrees).
T-42 minutes	Arm launch escape system. Launch vehicle power transfer test, LM switch to internal power.

T-37 minutes	Final launch vehicle range safety checks (to 35 minutes).
T-30 minutes	Launch vehicle power transfer test, LM switch over to internal power.
T-20 minutes to T-10 minutes	Shutdown LM operational instrumentation.
T-15 minutes	Spacecraft to full internal power.
T-6 minutes	Space vehicle final status checks.
T-5 minutes, 30 seconds	Arm destruct system.
T-5 minutes	Apollo access arm fully retracted.
T-3 minutes, 6 seconds	Firing command (automatic sequence).
T-50 seconds	Launch vehicle transfer to internal power.
T-8.9 seconds	Ignition start.
T-2 seconds	All engines running.
T-0	Liftoff.

NOTE: Some changes in the countdown are possible as a result of experience gained in the countdown test which occurs about two weeks before launch.

- more -

LAUNCH WINDOWS

The mission planning considerations for the launch phase of a lunar mission are, to a major extent, related to launch windows. Launch windows are defined for two different time periods: a "daily window" has a duration of a few hours during a given 24-hour period; a "monthly window" consists of a day or days which meet the mission operational constraints during a given month or lunar cycle.

Launch windows are based on flight azimuth limits of 72° to 100° (Earth-fixed heading east of north of the launch vehicle at end of the roll program), on booster and spacecraft performance, on insertion tracking, and on Sun elevation angle at the lunar landing site.

LAUNCH DATE	Launch Windows* OPEN	CLOSE	SUN ELEVATION ANGLE
April 16, 1972	12:54 pm	4:43 pm	11.9°
May 14, 1972	12:17 pm	4:01 pm	6.8°
May 15, 1972	12:30 pm	4:09 pm	6.8°
May 16, 1972	12:38 pm	4:13 pm	18.6°
June 13, 1972**	10:50 am	2:23 pm	13.0°
June 14, 1972**	10:57 am	2:26 pm	13.0°
June 15, 1972**	(------ To Be Determined ------)		

* April times are Eastern Standard Time; all others are Eastern Daylight Time.

** Launch window times and Sun elevation angles for June will be refined. The June 15, 1972 launch opportunity is under review.

- more -

GROUND ELAPSED TIME UPDATE

It is planned to update, if necessary, the actual ground elapsed time (GET) during the mission to allow the major flight plan events to occur at the pre-planned GET regardless of either a late liftoff or trajectory dispersions that would otherwise have changed the event times.

For example, if the flight plan calls for descent orbit insertion (DOI) to occur at GET 82 hours, 40 minutes and the flight time to the Moon is two minutes longer than planned due to trajectory dispersions at translunar injection, the GET clock will be turned back two minutes during the trans-lunar coast period so that DOI occurs at the pre-planned time rather than at 82 hours, 42 minutes. It follows that the other major mission events would then also be accomplished at the pre-planned times.

Updating the GET clock will accomplish in one adjustment what would otherwise require separate time adjustments for each event. By updating the GET clock, the astronauts and ground flight control personnel will be relieved of the burden of changing their checklists, flight plans, etc.

The planned times in the mission for updating GET will be kept to a minimum and will, generally, be limited to three updates. If required, they will occur at about 53, 97, and 150 hours into the mission. Both the actual GET and the update GET will be maintained in the MCC throughout the mission.

LAUNCH AND MISSION PROFILE

The Saturn V launch vehicle (SA-511) will boost the Apollo 16 spacecraft from Launch Complex 39A at the Kennedy Space Center, Fla., at 12:54 p.m. EST April 16, 1972, on an azimuth of 72 degrees.

The first stage (S-IC) will lift the vehicle 68.4 kilometers (37 nautical miles) above the Earth. After separation the booster stage will fall into the Atlantic Ocean about 668 km (361 nautical mi.) downrange from Cape Kennedy, approximately nine minutes, 14 seconds after liftoff.

The second stage (S-II) will push the vehicle to an altitude of about 173 km (93.6 nautical mi.). After separation the S-II stage will follow a ballistic trajectory which will plunge it into the Atlantic about 4,210 km (2,273 nautical mi.) downrange about 19 minutes, 51 seconds into the mission.

The single engine of the third stage (S-IVB) will insert the vehicle and spacecraft into a 173-km (93-nautical mi.) circular Earth parking orbit before it is cut off for a coast period. When reignited, the engine will inject the Apollo spacecraft into a trans-lunar trajectory.

APOLLO 16
FLIGHT PROFILE

SUBSATELLITE EJECTION (REV 73)

TRANSEARTH INJECTION (REV 76)

CSM 60 NM

RENDEZVOUS

BRAKING (REV 51)

CSM PLANE CHANGES (REVS 40 & 51)

CSM ORBIT CIRCULARIZATION 52 x 68 NM (REV 12)

CSM/LM SEPARATION (REV 12)

CSM/LM DOI 9 x 60 NM (REV 2)

CSM/LM LOI ~60 x 170 NM

CSM ORBIT CHANGE 55 x 85 NM (REV 73)

IN-FLIGHT EVA

CSM TRANSEARTH TRAJECTORY

LM/CSM DOCKING (REV 51)

LM ASCENT (REV 50)

ASCENT STAGE JETTISON (REV 53)

LM LANDING

LM PDI (REV 13)

SIM DOOR JETTISON

S/C TRANSLUNAR TRAJECTORY

S-IVB IMPACT TRAJECTORY

90-NM EARTH PARKING ORBIT

LAUNCH

EARTH ORBIT INSERTION

S-IVB APS EVASIVE MANEUVER

CM/SM SEPARATION

CM LANDING AND RECOVERY

S-IVB 2ND BURN CUTOFF TRANSLUNAR INJECTION (TLI)

S/C SEPARATION, TRANSPOSITION, DOCKING & EJECTION

4174

APOLLO 16 vs APOLLO 15
OPERATIONAL DIFFERENCES

ITEM	APOLLO 16	APOLLO 15
LAUNCH AZIMUTH	72 - 100°	80 - 100°
INCLINATION	9°	26°
DOI TRIM MANEUVER	NOT PLANNED	ACCOMPLISHED
HIGH ALTITUDE LANDMARK TRACKING AND TV OF LANDING SITE	NO	YES
PDI	REV 13	REV 14
SURFACE STAY TIME	~73 HOURS	66.9 HOURS
SURFACE REST CYCLE	8 HOURS (REST BEFORE LIFTOFF)	7 - 7 1/2 HOURS (NO REST BEFORE LIFTOFF)
EVA's: DURATION (HRS) (PLANNED)	7 - 7 - 7	7 - 7 - 6
SEVA	NO	YES
TRAVERSE STATION TIMES	9:40 (PLANNED)	7:53 (PLANNED) 5:00 (ACTUAL)
SURFACE ACTIVITIES	ALSEP DEPLOYED FIRST	TRAVERSE FIRST
LUNAR ORBIT PLANE CHANGES	2	1
EARTH RETURN INCLINATION	62°	40°
LANDING	PLAN TO RECOVER PARA-CHUTES AND RETAIN RCS PROPELLANTS ON BOARD	NEITHER APOLLO 16 TYPE EVENTS WERE PLANNED

COMPARISON OF APOLLO MISSIONS

	PAYLOAD DELIVERED TO LUNAR SURFACE KG (LBS)	EVA DURATION (HR:MIN)	SURFACE DISTANCE TRAVERSED (KM)	SAMPLES RETURNED KG (LBS)
APOLLO 11	104 (225)	2:24	.25	20.7 (46)
APOLLO 12	166 (365)	7:29	2.0	34.1 (75)
APOLLO 14	209 (460)	9:23	3.3	42.8 (94)
APOLLO 15	550 (1210)	18:33	27.9	76.6 (169)
APOLLO 16 (PLANNED)	558 (1228)	21:00	25.7	88.6 (195)

LAUNCH EVENTS

Time Hrs	Min	Sec	Event	Vehicle Wt Kilograms (Pounds)*	Altitude Meters (Feet)*	Velocity Mtrs/Sec (Ft/Sec)*	Range Kilometers (Naut Mi)*
00	00	00	First Motion	2,920,956 (6,439,605)	60 (198)	0 (0)	0 (0)
00	01	20	Maximum Dynamic Pressure	1,853,350 (4,085,937)	12,646 (41,491)	481 (1,577)	5 (3)
00	02	18	S-IC Center Engine Cutoff	1,084,409 (2,390,712)	47,153 (154,703)	1,712 (5,618)	51 (27)
00	02	40	S-IC Outboard Engines Cutoff	843,020 (1,858,541)	67,371 (221,034)	2,369 (7,773)	91 (49)
00	02	42	S-IC/S-II Separation	676,542 (1,491,519)	69,094 (226,685)	2,375 (7,792)	95 (51)
00	02	44	S-II Ignition	676,542 (1,491,519)	70,645 (231,775)	2,369 (7,771)	98 (53)
00	03	12	S-II Aft Interstage Jettison	644,029 (1,419,840)	95,024 (311,758)	2,474 (8,117)	161 (87)
00	03	18	Launch Escape Tower Jettison	633,419 (1,396,451)	99,428 (326,207)	2,503 (8,213)	174 (94)
00	07	40	S-II Center Engine Cutoff	303,470 (669,037)	174,389 (572,142)	5,201 (17,065)	1,095 (591)
00	09	17	S-II Outboard Engines Cutoff	216,673 (477,682)	174,130 (571,291)	6,568 (21,550)	1,650 (891)

(continued)

* English measurements given in parentheses.

Time Hrs Min Sec	Event	Vehicle Wt Kilograms (Pounds)*	Altitude Meters (Feet)*	Velocity Mtrs/Sec (Ft/Sec)*	Range Kilometers (Naut Mi)*
00 09 18	S-II/S-IVB Separation	170,989 (376,967)	174,161 (571,395)	6,571 (21,558)	1,657 (895)
00 09 21	S-IVB First Ignition	170,947 (376,874)	174,235 (571,637)	6,571 (21,559)	1,677 (905)
00 11 44	S-IVB First Cutoff	140,103 (308,873)	172,908 (567,282)	7,400 (24,279)	2,643 (1,427)
00 11 54	Parking Orbit Insertion	140,039 (308,734)	172,914 (567,302)	7,402 (24,284)	2,715 (1,466)
02 33 35	S-IVB Second Ignition	139,052 (306,557)	175,919 (577,162)	7,404 (24,292)	16,203 (8,749)
02 39 19	S-IVB Second Cutoff	65,620 (144,666)	305,315 (1,001,688)	10,441 (34,256)	13,320 (7,192)
02 39 29	Trans-Lunar Injection	65,551 (144,514)	319,199 (1,047,241)	10,433 (34,230)	13,222 (7,139)

* English measurements given in parentheses.

ACTIVITY	TIME	PURPOSE
● VISUAL LIGHT FLASH PHENOMENON.		TO STUDY THE ORIGIN OF VISUAL LIGHT FLASHES OBSERVED BY ASTRONAUTS ON PREVIOUS MISSIONS.
● EYE SHIELDS	TRANSLUNAR & TRANSEARTH	
● ALFMED	TRANSLUNAR	
● ELECTROPHORETIC SEPARATION	TRANSLUNAR	TO STUDY A PROCESS OF MATERIALS PURIFICATION WHICH MAY PERMIT PRO-DUCTION SEPARATION OF MATERIALS OF HIGHER PURITY THAN CAN CURRENTLY BE PRODUCED ON EARTH.

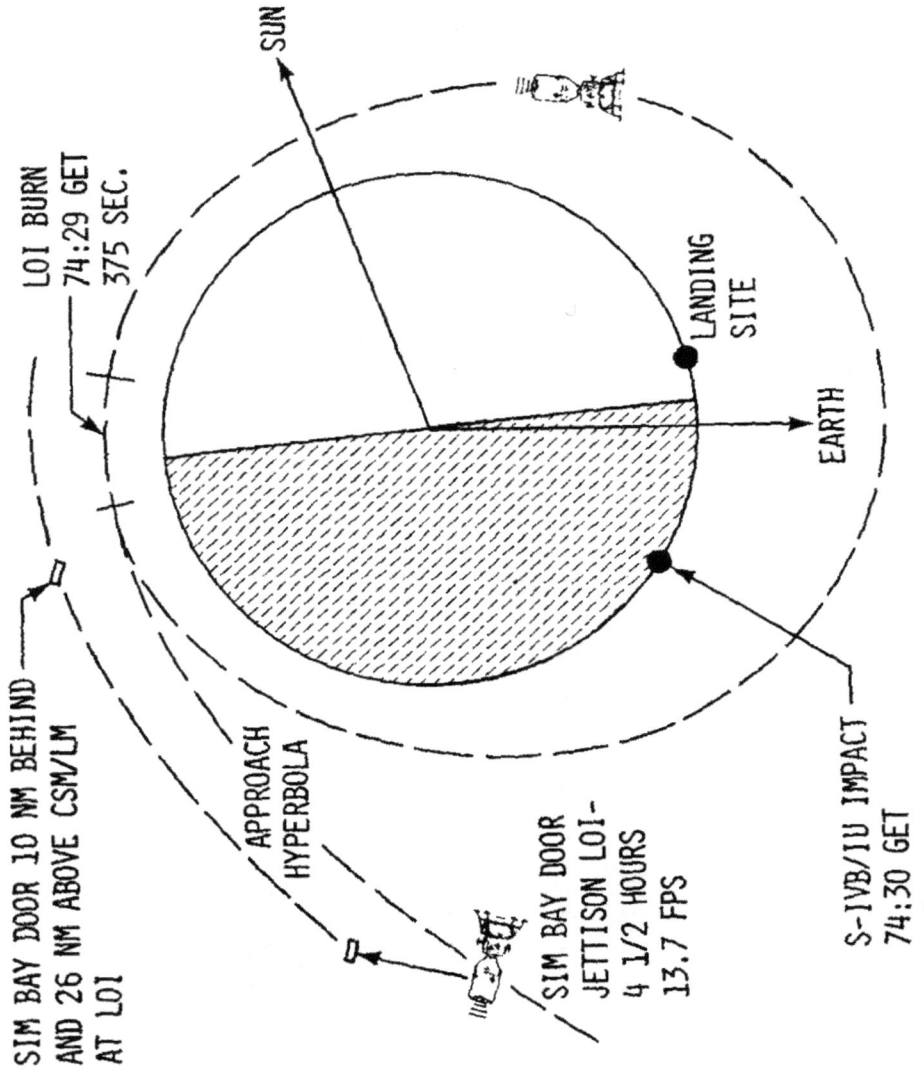

APOLLO 16

SUN

LOI BURN
74:29 GET
375 SEC.

SIM BAY DOOR 10 NM BEHIND
AND 26 NM ABOVE CSM/LM
AT LOI

APPROACH
HYPERBOLA

SIM BAY DOOR
JETTISON LOI-
4 1/2 HOURS
13.7 FPS

LANDING
SITE

EARTH

S-IVB/IU IMPACT
74:30 GET

MISSION EVENTS

Events	GET hrs:min	Date/EST	Velocity change m/sec (ft/sec)		Purpose and resultant orbit
Translunar injection (S-IVB engine start)	2:33	16/3:27 pm	3053	(10,018)	Injection into translunar trajectory with 13 km (71 nm) pericynthion
CSM separation, docking	3:04	16/3:58 pm	---		Mating of CSM and LM
Ejection from SLA	3:59	16/4:53 pm	.3	(1)	Separates CSM-LM from S-IVB/SLA
S-IVB evasive maneuver	4:23	16/5:17 pm	3	(9.8)	Provides separation prior to S-IVB propellant dump and thruster maneuver to cause lunar impact
S-IVB residual propellant dump	4:44	16/6:38 pm			
APS impact burn	5:30	16/7:24 pm			
APS correction burn	9:30	16/11:24 pm			
Midcourse correction 1	TLI+9 hr	17/00:33 am	0*		*These midcourse corrections have a nominal velocity change of 0 m/sec, but will be calculated in real time to correct TLI dispersions; Trajectory remains within capability of a docked-DPS TEI burn should SPS fail to ignite.
Midcourse correction 2	TLI+28 hr	17/7:33 pm	0*		
Midcourse correction 3	LOI-22 hr	18/5:23 pm	0*		
Midcourse correction 4	LOI-5 hr	19/10:23 am	0*		
SIM door jettison	LOI-4.5 hr	19/10:53 am	4.2	(13.7)	
Lunar Orbit insertion	74:29	19/3:23 pm	-855.6	(-2807)	Inserts Apollo 16 into 108x316 km (58x170 nm) elliptical lunar orbit.
S-IVB impacts lunar surface	74:30	19/3:24 pm			Seismic event for Apollo 12, 14 and 15 passive seismometers. Target: 2.2 degrees south latitude by 21.4 degrees west longitude.
Descent orbit insertion	78:36	19/7:30 pm	-62.8	(206)	SPS burn places CSM/LM into 20x109 km (11x59 nm) lunar orbit.

- more -

Events	GET hrs:min	Date/EST	Velocity change m/sec (ft/sec)	Purpose and resultant orbit
CSM/LM undocking	96:14	20/01:08 pm	---	
CSM circularization burn	97:42	20/2:36 pm	30.3 (99.6)	Inserts CSM into 96x111 km (52x60 nm) orbit (SPS burn)
LM powered descent	98:35	20/3:29 pm	-2041 (6696)	Three-phase DPS burn to brake LM out of transfer orbit, vertical descent and lunar surface touchdown.
LM lunar surface touchdown	98:46	20/3:41 pm	---	Lunar exploration, deploy ALSEP, collect geological samples, photography.
EVA-1 begins	102:25	20/7:19 pm	---	See separate EVA timelines
EVA-2 begins	124:50	21/5:44 pm	---	See separate EVA timelines
EVA-3 begins	148:25	22/5:19 pm	---	See separate EVA timelines
CSM plane change 1	152:29	22/9:23 pm	48 (159)	Changes CSM orbital plane by 1.7 degree to coincide with LM orbital plane at time of LM ascent from surface
LM ascent	171:45	23/4:39 pm	1,843 (6048)	Boosts ascent stage into 16.6x84 km (9x45.4 nm) lunar orbit for rendezvous with CSM
Lunar orbit insertion	171:52	23/4:46 pm		Boosts ascent stage into 81.4x114.5 km (44x61.9 nm) catch-up orbit; LM trails CSM by 59.2 km (32 nm) and 27.7 km (15 nm) below at TPI burn time
Terminal phase initiate (TPI) LM APS	172:39	23/5:33 pm	15 50	
Braking: 4 LM RCS burns	173:20	23/6:14 pm	10 33	Line-of-sight terminal phase braking to place LM in 110.7x109.8 km (59.8x59.3 nm) orbit for final approach, docking.
Docking	173:40	23/6:34 pm	---	CDR and LMP transfer back to CSM
LM jettison, separation	177:31	23/10:25 pm	---	Prevents recontact of CSM with LM ascent stage for remainder of mission

- more -

POWERED DESCENT VEHICLE POSITIONS

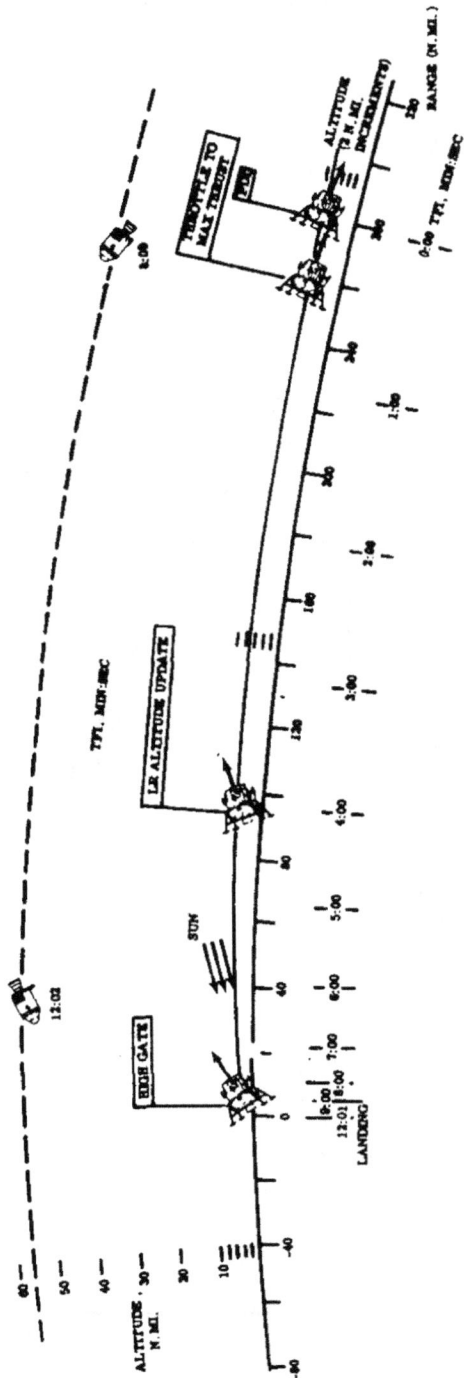

SUMMARY

EVENT	TFI MIN:SEC	V_I, FPS	$\dot{H}$ FPS	H FT	ΔV FPS
POWERED DESCENT INITIATION	0:00	5555	-5	52700	0
THROTTLE TO MAXIMUM THRUST	0:26	5527	-4	52600	28
YAW TO VERTICAL	3:00	4096	-56	46840	1476
THROTTLE RECOVERY	7:20	1188	-91	24740	4559
HIGH GATE	9:20	342	-174	7970	5602
LOW GATE	10:42	76	-21	583	6246
LANDING	12:01.5	0	-5	0	6697

APPROACH PHASE

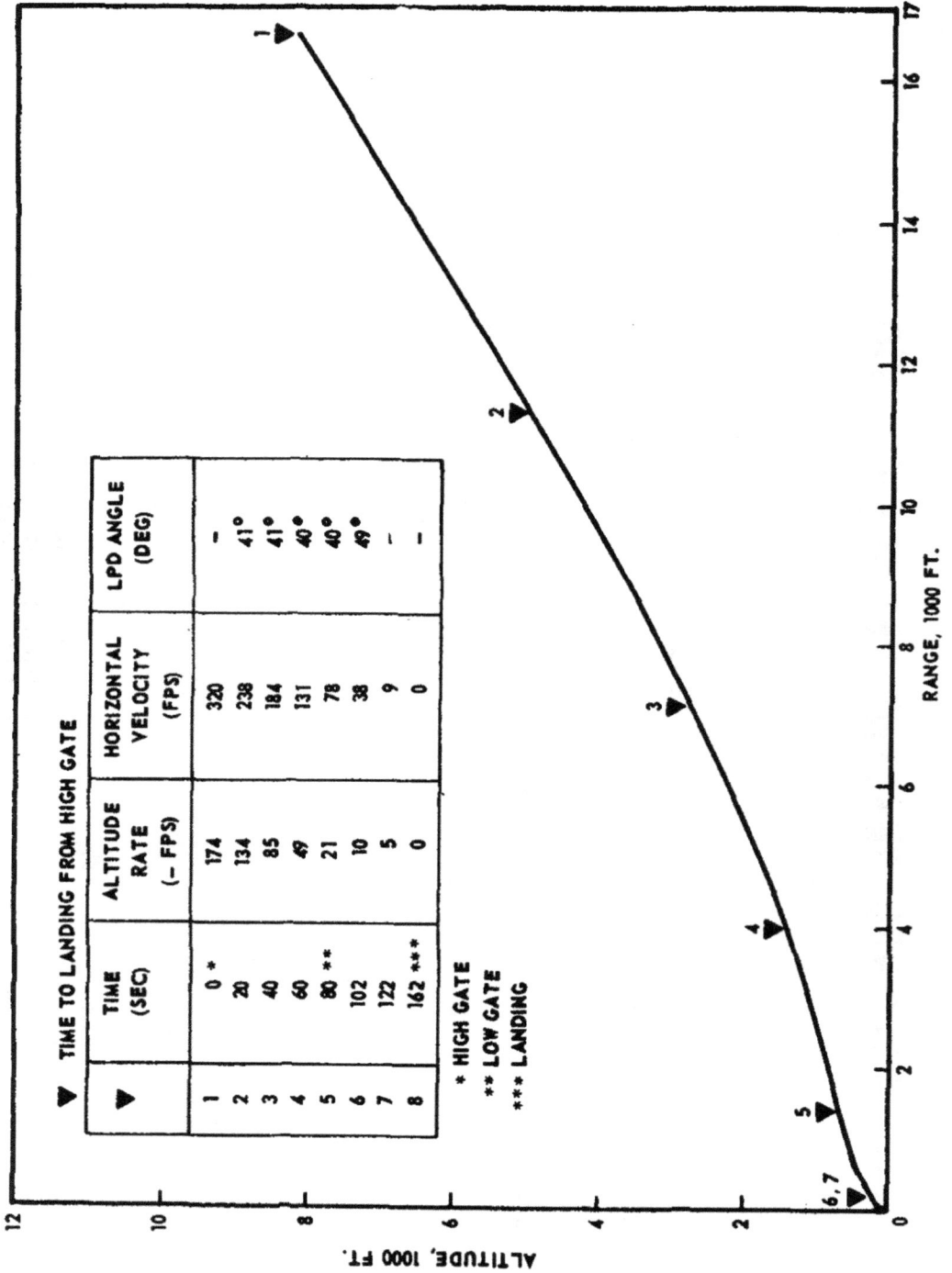

▼ TIME TO LANDING FROM HIGH GATE

▼	TIME (SEC)	ALTITUDE RATE (– FPS)	HORIZONTAL VELOCITY (FPS)	LPD ANGLE (DEG)
1	0 *	174	320	–
2	20	134	238	41°
3	40	85	184	41°
4	60	49	131	40°
5	80 **	21	78	40°
6	102	10	38	49°
7	122	5	9	–
8	162 ***	0	0	–

* HIGH GATE
** LOW GATE
*** LANDING

ALTITUDE, 1000 FT.

RANGE, 1000 FT.

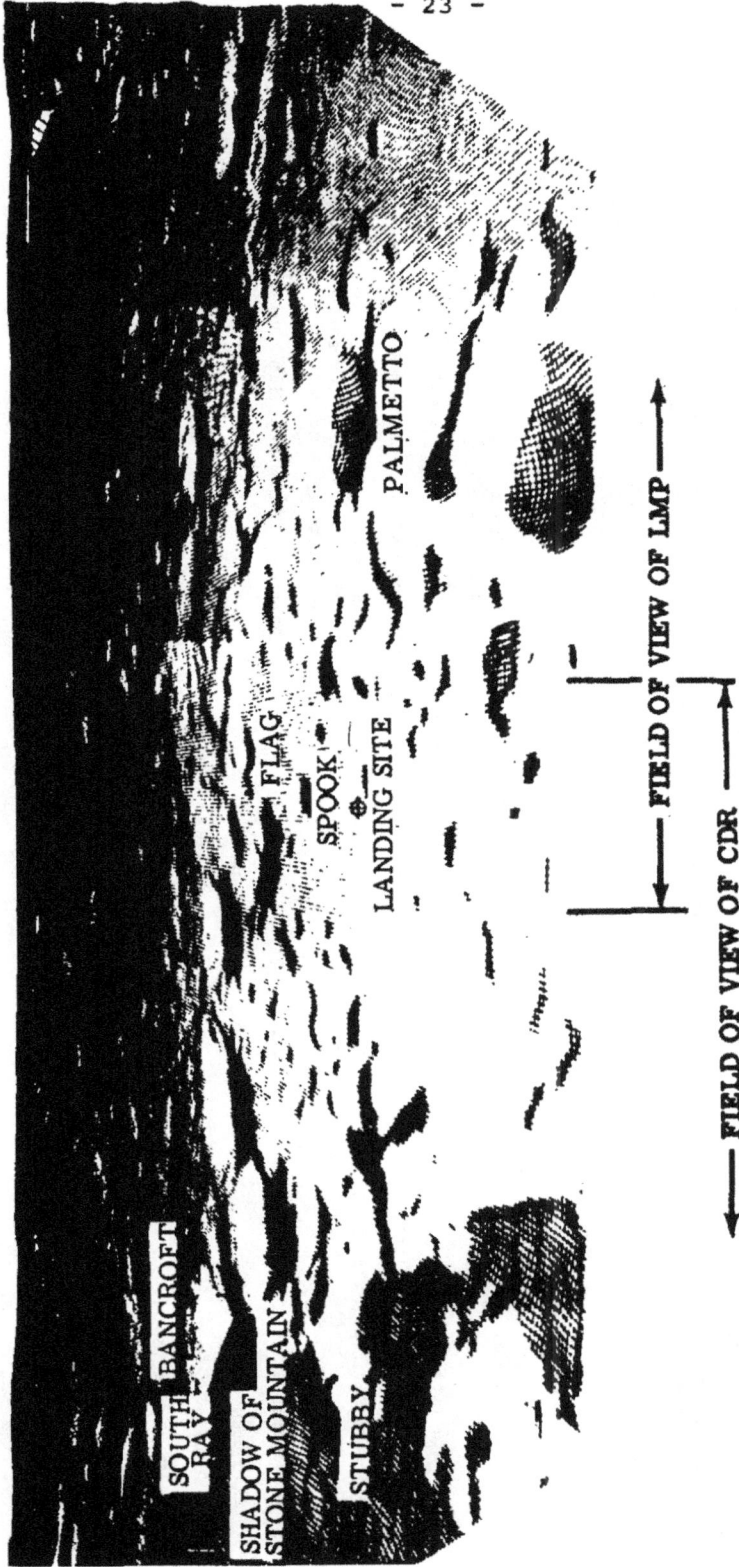

FIRST VIEW OF DESCARTES
(HIGH GATE + 8 SECONDS)

Events	GET hrs:min	Date/EST	Velocity change m/sec (ft/sec)		Purpose and resultant orbit
LM ascent stage deorbit (RCS burn)	179:16	24/00:10 am	-70	(-229)	ALSEP seismometers at Apollo 16, 15, 14 and 12 landing sites record impact.
LM impact	179:39	24/00:33 am	---		Impact at about 1691 m/sec (5550 fps) at -3.2 degree angle. 23 km (12 nm) west of Apollo 16 ALSEP.
CSM plane change 2	193:14	24/2:08 pm	86	(282)	Increase lunar surface photo coverage, changes plane 3 degrees.
CSM shaping burn	216:49	25/1:43 pm	11	(38)	Adjusts CSM orbit for later subsatellite jettison orbit: 102x147 km (55x85 nm).
Subsatellite jettison	218:02	25/2:56 pm	---		Lunar orbit science experiments
Transearth injection (TEI)	222:21	25/7:15 pm	979	(3212)	Inject CSM into transearth trajectory
Midcourse correction 5	239:21	26/12:17 pm	0		Transearth midcourse corrections will be computed in real time for entry corridor control and recovery area weather avoidance.
Transearth EVA	241:57	26/2:51 pm	---		Retrieve SM SIM bay film canisters.
Midcourse correction 6	EI-22 hrs	27/5:17 pm	0		
Midcourse correction 7	EI-3 hrs	28/12:17 pm	0		
CM/SM separation	290:08	28/3:02 pm	---		Command module oriented for Earth atmosphere entry
Entry interface (121.9 km, 400,000 ft)	290:23	28/3:17 pm			Command module enters Earth's atmosphere at 11,026 m/s (36,175 fps).
Splashdown	290:36	28/3:30 pm			Landing 2111 km (1140 nm) downrange from entry; splash at 5 degrees north latitude by 158 degrees west longitude.

- more -

FILM RETRIEVAL FROM THE SIM BAY

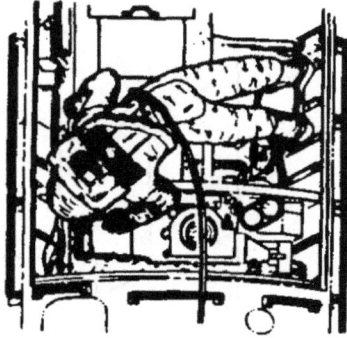

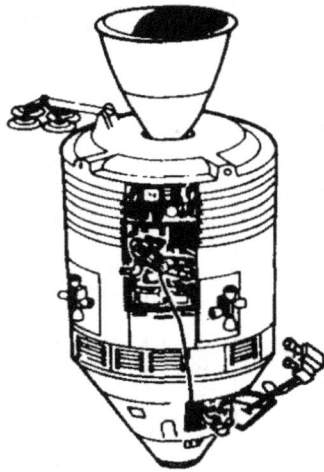

EVA TIME LINE

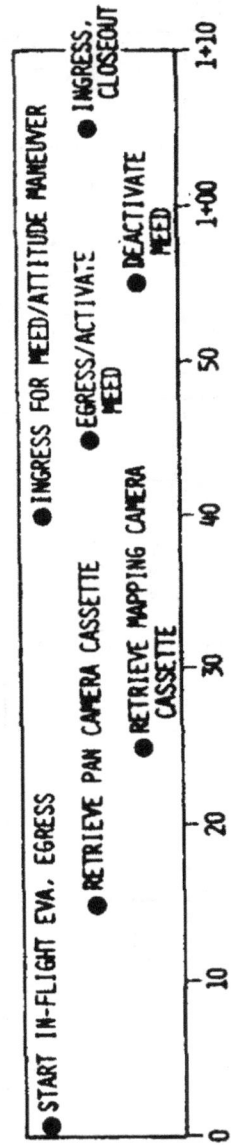

- START IN-FLIGHT EVA, EGRESS
- RETRIEVE PAN CAMERA CASSETTE
- RETRIEVE MAPPING CAMERA CASSETTE
- INGRESS FOR MEED/ATTITUDE MANEUVER
- EGRESS/ACTIVATE MEED
- DEACTIVATE MEED
- INGRESS, CLOSEOUT

0   10   20   30   40   50   1+00   1+10

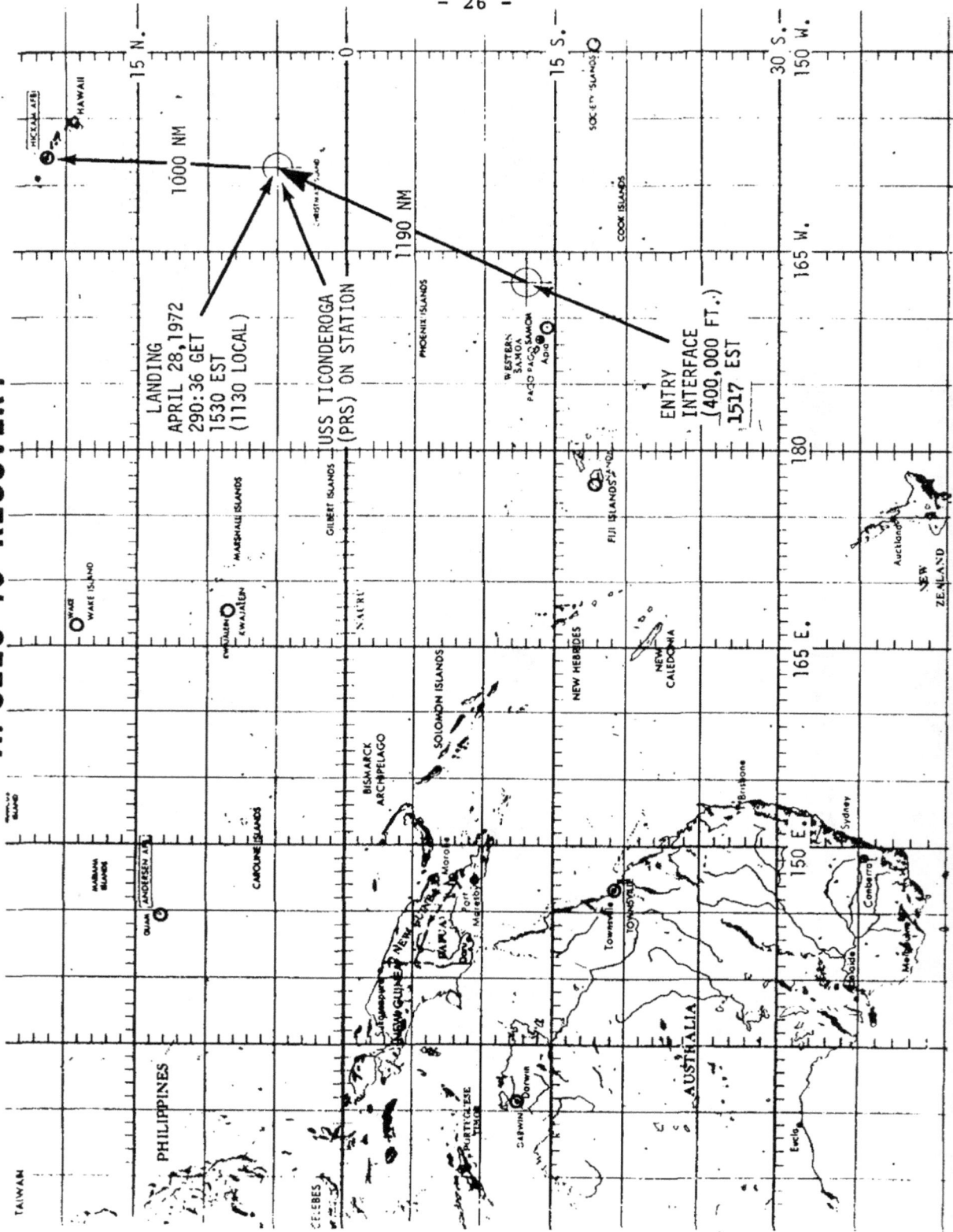

APOLLO 16 RECOVERY

APOLLO 16

CREW POST-LANDING ACTIVITIES

DAYS FROM RECOVERY	DATE	ACTIVITY
SPLASHDOWN	APRIL 28	
R + 1	APRIL 29	DEPART SHIP, ARRIVE HAWAII
R + 2	APRIL 30	DEPART HAWAII, ARRIVE HOUSTON
R + 3 THRU R + 15	MAY 1-12	CREW DEBRIEFING PERIOD

APOLLO 16 ALTERNATE MISSIONS

● EARTH ORBIT MISSION (NO TLI)

   ● PHOTOGRAPHIC MISSION OVER U.S. OF ABOUT 6 DAYS DURATION, X-RAY AND GAMMA RAY SPECTROMETERS WILL ALSO BE OPERATED, CMP EVA TO RETRIEVE FILM CASSETTES.

● LUNAR ORBIT MISSIONS

   ● APPROXIMATELY 6 DAYS IN LUNAR ORBIT

   ● USE OF SIM BAY EXPERIMENTS

   ● SUBSATELLITE JETTISON REQUIRED

      TYPES OF CONTINGENCIES

        ● CSM ALONE

        ● CSM/LM (OPERABLE AND INOPERABLE DPS CASES)

        ● FROM LM ABORTS

EVA MISSION EVENTS

Events	GET hrs:min	Date/EST
Depressurize LM for EVA 1	102:25	Apr 20/7:19 pm
CDR steps onto surface	102:40	20/7:34 pm
LMP steps onto surface	102:45	20/7:39 pm
Crew offloads LRV	102:53	20/7:47 pm
CDR test drives LRV	103:13	20/8:07 pm
LRV parked near MESA	103:15	20/8:09 pm
CDR offloads, deploys far-UV camera/ spectroscope	103:24	20/8:18 pm
LMP mounts LCRU, TV on LRV	103:25	20/8:19 pm
Crew loads geology equipment on LRV	103:37	20/8:31 pm
CDR deploys United States flag	103:50	20/8:44 pm
LMP carries ALSEP "barbell" to deployment site	104:02	20/8:56 pm
CDR drives LRV to ALSEP site	104:12	20/9:06 pm
Crew begins ALSEP deploy	104:20	20/9:14 pm
ALSEP deploy complete	106:22	20/11:16 pm
Crew drives to station 1 (Flag Crater)	106:27	20/11:21 pm
Crew arrives at station 1 for rake/ soil sample, crater sample	106:39	20/11:33 pm
Crew drives to station 2 (Spook crater)	107:22	21/00:06 am
Crew arrives at station 2 for crater area sampling, magnetometer site measurement	107:27	21/00:21 am
Crew returns to LM/ALSEP area Station 3	108:23	Apr 21/1:17 am

- more -

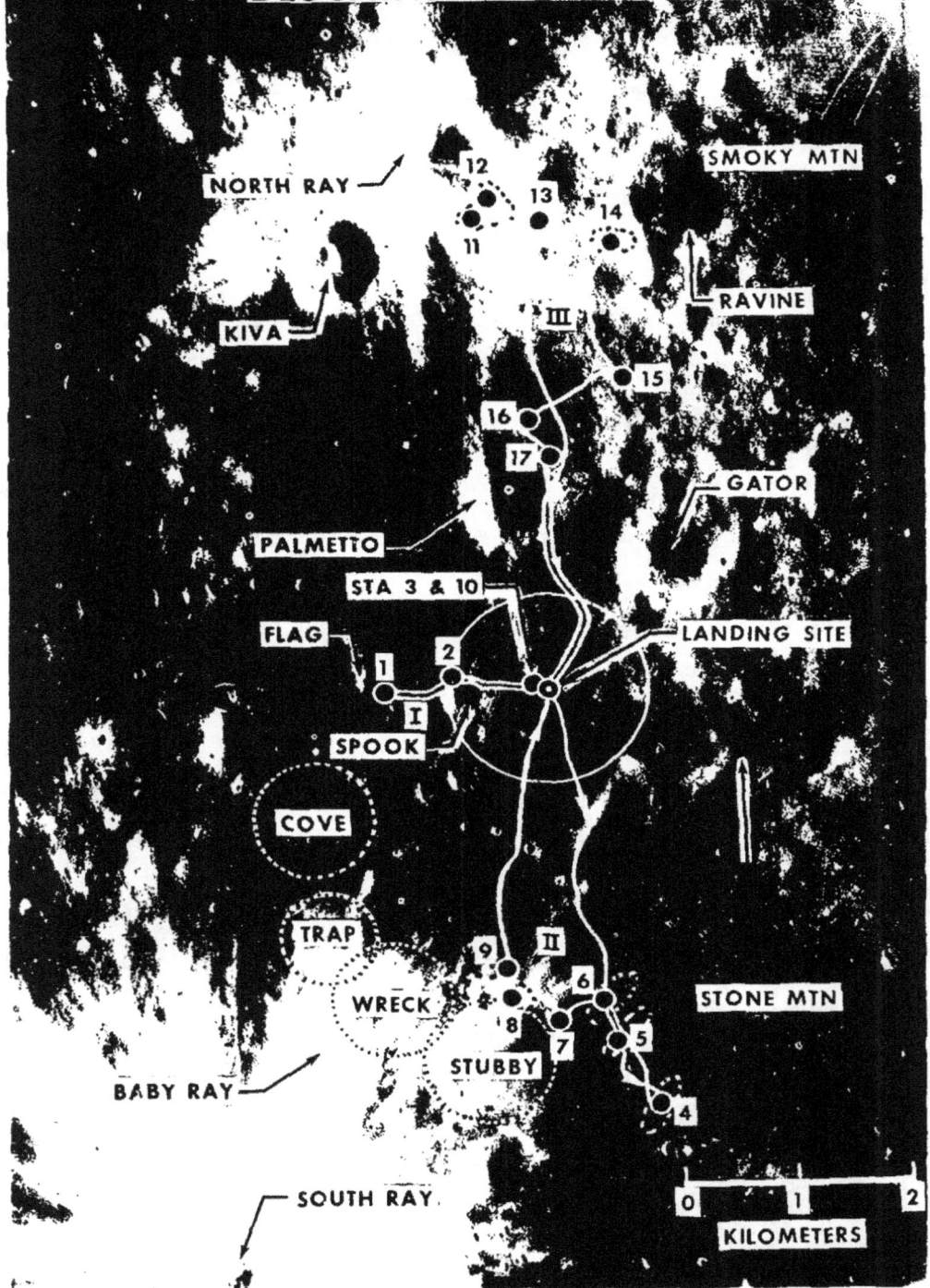

DESCARTES LRV TRAVERSES

APOLLO 16 TRAVERSE COMPARISON WITH APOLLO 15

	DURATION (HR:MIN)		TRAVERSE STATION TIME (HR:MIN)		LM/ALSEP AREA TIME (HR:MIN)		LRV DRIVING TIME (HR:MIN)		TRAVERSE DISTANCE (KM)	
	PLAN	ACTUAL	PLAN	ACTUAL	PLAN	ACTUAL	PLAN	ACTUAL	PLAN	ACTUAL
EVA I APOLLO 15	7:00	6:32	1:16	1:10	4:35	4:14	1:08	1:08	8.2	10.3
APOLLO 16	7:00	-	1:53	-	4:41	-	:26	-	3.2	-
EVA II APOLLO 15	7:00	7:12	3:32	2:37	1:29	3:11	1:59	1:24	14.3	12.5
APOLLO 16	7:00	-	4:11	-	1:30	-	1:19	-	9.5	-
EVA III APOLLO 15	6:00	4:49	3:05	1:23	1:27	2:50	1:28	:36	10.5	5.1
APOLLO 16	7:00	-	3:36	-	1:40	-	1:44	-	12.6	-
TOTALS APOLLO 15	20:00	18:33	7:53	5:00	7:31	10:15	4:35	3:08	33.0	27.9
APOLLO 16	21:00	-	9:40	-	7:51	-	3:29	-	25.2	-

APOLLO 16 LUNAR SURFACE TIMELINE

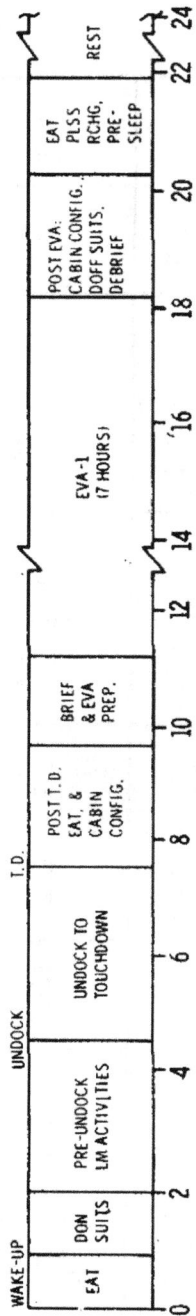

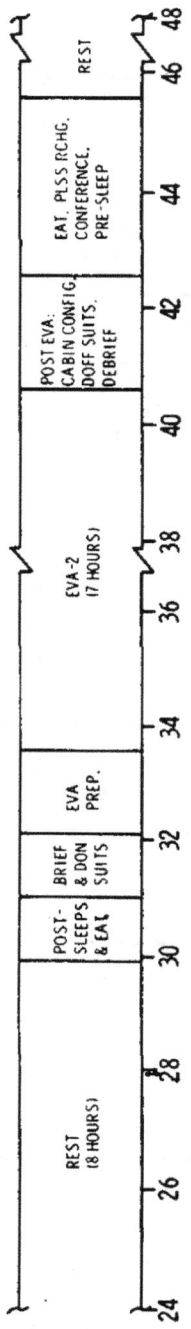

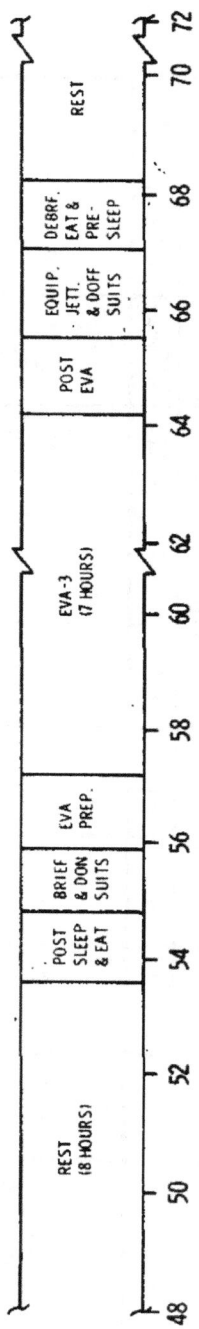

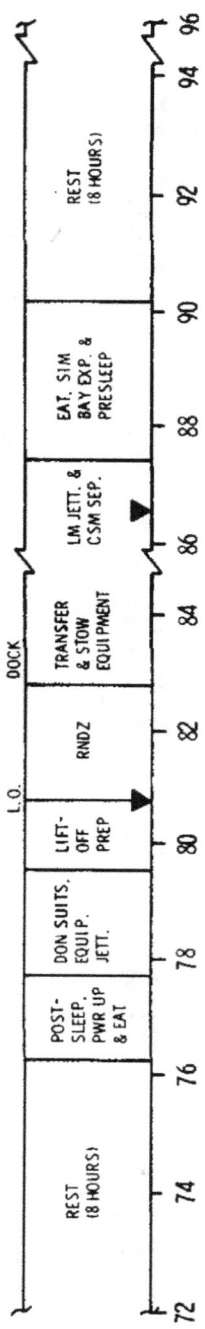

Row 1 (0–24):

WAKE-UP | UNDOCK | T.D.

EAT | DON SUITS | PRE-UNDOCK LM ACTIVITIES | UNDOCK TO TOUCHDOWN | POST T.D. EAT, & CABIN CONFIG. | BRIEF & EVA PREP. | EVA-1 (7 HOURS) | POST EVA. CABIN CONFIG., DOFF SUITS, DEBRIEF | EAT, PLSS RCHG, PRE-SLEEP | REST

0 2 4 6 8 10 12 14 16 18 20 22 24

Row 2 (24–48):

REST (8 HOURS) | POST-SLEEPS & EAT | BRIEF & DON SUITS | EVA PREP. | EVA-2 (7 HOURS) | POST EVA. CABIN CONFIG, DOFF SUITS, DEBRIEF | EAT, PLSS RCHG, CONFERENCE, PRE-SLEEP | REST

24 26 28 30 32 34 36 38 40 42 44 46 48

Row 3 (48–72):

REST (8 HOURS) | POST SLEEP & EAT | BRIEF & DON SUITS | EVA PREP. | EVA-3 (7 HOURS) | POST EVA | EQUIP. JETT. & DOFF SUITS | DEBRF. EAT & PRE-SLEEP | REST

48 50 52 54 56 58 60 62 64 66 68 70 72

Row 4 (72–96):

L.O. | DOCK

REST (8 HOURS) | POST-SLEEP, PWR UP & EAT | DON SUITS, EQUIP. JETT. | LIFT-OFF PREP | RNDZ | TRANSFER & STOW EQUIPMENT | LM JETT. & CSM SEP. | EAT, SIM BAY EXP. & PRESLEEP | REST (8 HOURS)

72 74 76 78 80 82 84 86 88 90 92 94 96

- 33 -

SUMMARY TIME LINE
FROM ALSEP OFFLOAD THROUGH ALSEP DEPLOYMENT

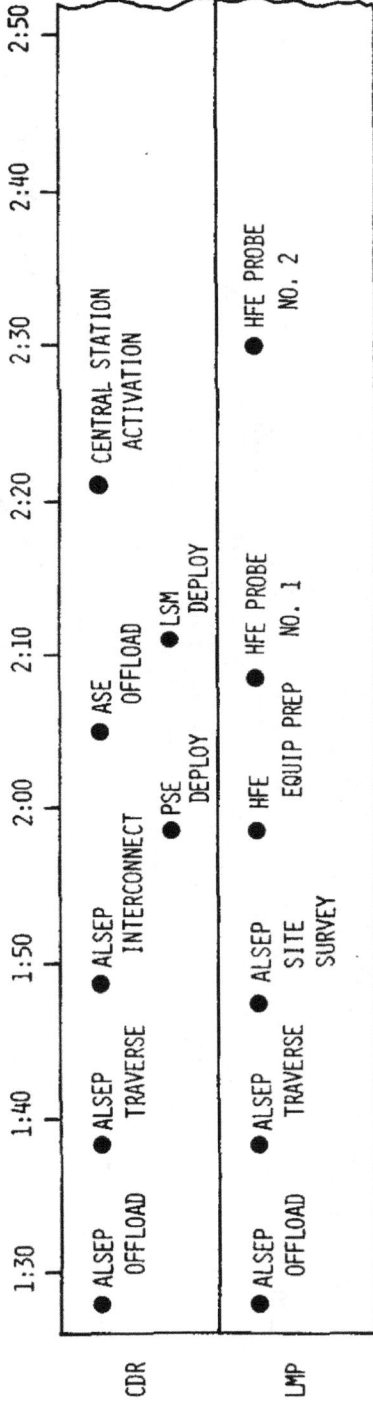

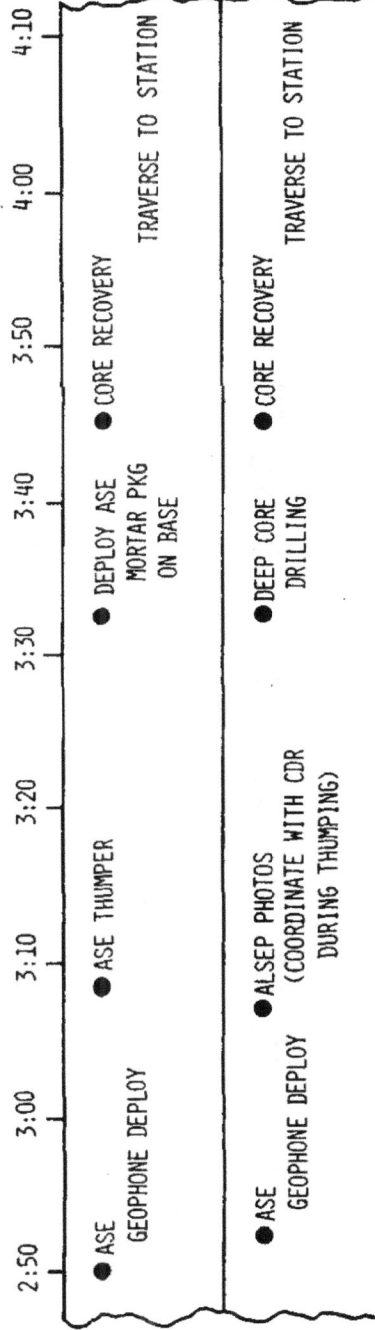

CDR

1:30	1:40	1:50	2:00	2:10	2:20	2:30	2:40	2:50

- ALSEP OFFLOAD
- ALSEP TRAVERSE
- ALSEP INTERCONNECT
- ASE OFFLOAD
- CENTRAL STATION ACTIVATION

- ALSEP OFFLOAD
- ALSEP TRAVERSE
- ALSEP SITE SURVEY
- PSE DEPLOY
- HFE EQUIP PREP
- HFE PROBE NO. 1
- LSM DEPLOY
- HFE PROBE NO. 2

LMP

2:50	3:00	3:10	3:20	3:30	3:40	3:50	4:00	4:10

- ASE GEOPHONE DEPLOY
- ASE THUMPER
- DEPLOY ASE MORTAR PKG ON BASE
- CORE RECOVERY
- TRAVERSE TO STATION

- ASE GEOPHONE DEPLOY
- ALSEP PHOTOS (COORDINATE WITH CDR DURING THUMPING)
- DEEP CORE DRILLING
- CORE RECOVERY
- TRAVERSE TO STATION

NEAR LM LUNAR SURFACE ACTIVITY

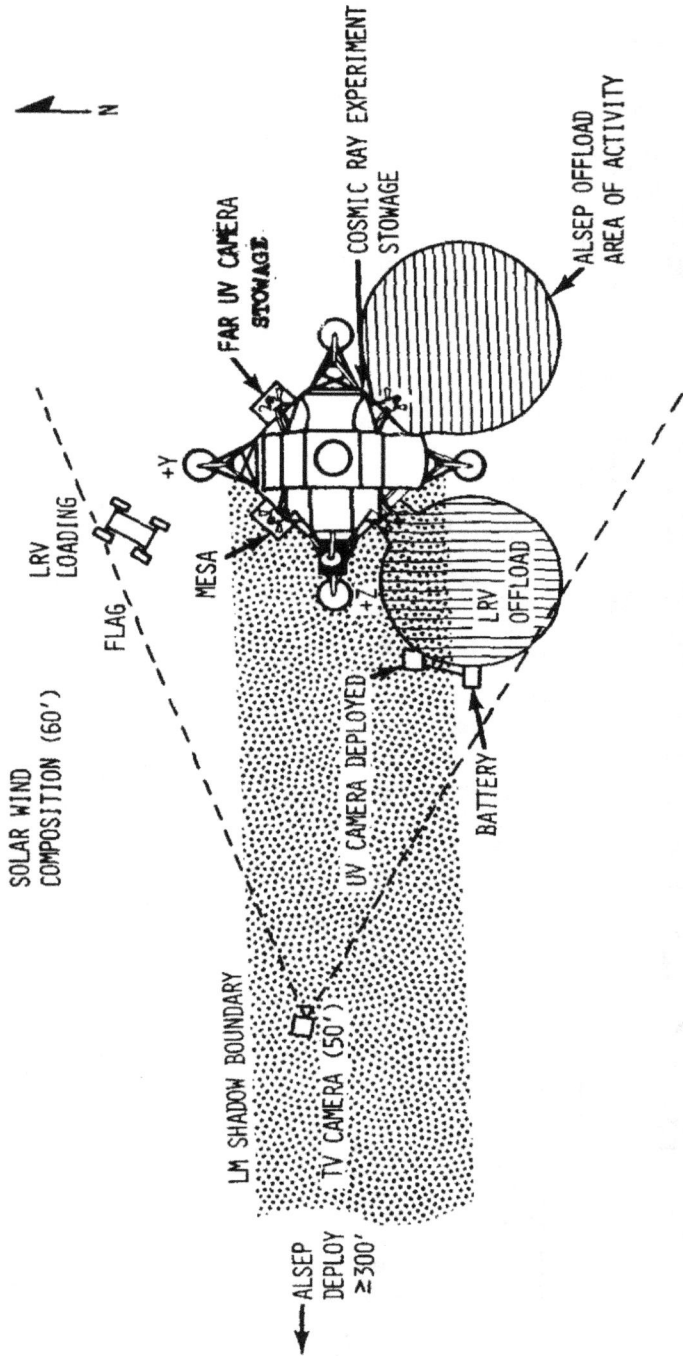

Events	GET hrs:min		Date/EST
Crew arrives at station 3 to retrieve deep core. Comprehensive sampling, arm mortar package, perform LRV "Gran Prix".	108:30		21/1:24 am
Crew returns to LM	108:44		21/1:38 am
Crew arrives at LM	108:45		21/1:39 am
LMP deploys solar wind composition experiment	109:02		21/1:56 am
CDR offloads geology equipment, film mags	109:05		21/1:59 am
LMP ingresses LM	109:15		21/2:09 am
CDR ingresses LM	109:22		21/2:16 am
Repressurize LM, end EVA 1	109:25		21/2:19 am
Depressurize LM for EVA 2	124:50	Apr	21/5:44 pm
CDR steps onto surface	125:10		21/5:59 pm
LMP steps onto surface	125:10		21/6:04 pm
Crew completes loading LRV for geology traverse	125:30		21/6:24 pm
Crewmen load geological gear on each other's PLSS	125:32		21/6:26 pm
Crew drives to station 4 (upper slope of Stone Mountain)	125:40		21/6:34 pm
Crew arrives at station 4 for rake soil, documented and double core samples, photo panorama, penetrometer reading	126:15		21/7:09 pm
Crew drives to station 5 (on slope of Stone Mountain)	127:1$		21/8:07 pm
Crew arrives at station 5 for documented samples, photo panorama	127:19		21/8:13 pm

- more -

Events	GET hrs:min	Date/EST
Crew drives to station 6 (at base of Stone Mountain)	127:59	Apr 21/8:53 pm
Crew arrives at station 6 for documented samples, photo panorama	128:02	21/8:56 pm
Crew drives to station 7 (Stubby Crater, base of Stone Mountain)	128:22	21/9:16 pm
Crew arrives at station 7 for documented samples, photo panorama	128:26	21/9:20 pm
Crew drives to station 8 (rays from South Ray Crater)	128:41	21/9:35 pm
Crew arrives at station 8 for rake soil, documented double core samples, photo panorama	128:46	21/9:40 pm
Crew drives to station 9 (Cayley Plains)	129:45	21/10:39 pm
Crew arrives at Station 9 for surface and core samples, photo panorama	129:50	21/10:44 pm
Crew drives to station 10 (LM/ALSEP area)	130:15	21/11:09 pm
Crew arrives at station 10 for soil mechanics investigation, double core, radial sample, photo panorama	130:36	21/11:30 pm
Crew drives back to LM	131:09	22/00:03 am
Crew arrives at LM	131:10	22/00:04 am
Crew unloads LRV, packs sample containers, clean-up	131:20	22/00:14 am
LMP ingresses LM	131:37	22/00:31 am
CDR ingresses LM	131:48	22/00:42 am
Repressurize LM, end EVA 2	131:50	22/00:44 am

- more -

Events	GET hrs:min		Date/EST
Depressurize LM for EVA 3	148:25	Apr	22/5:19 pm
CDR steps onto surface	148:40		22/5:34 pm
LMP steps onto surface	148:46		22/5:40 pm
Crew completes loading LRV, PLSSs for geology traverse	149:02		22/5:56 pm
Crew drives to station 11 (South rim, North Ray Crater)	149:10		22/6:04 pm
Crew arrives at station 11 for polari-metric photos, documented samples, photo panorama	149:54		22/6:48 pm
Crew drives to station 12 (North Ray, Crater, southeast rim, large block area)	150:49		22/7:43 pm
Crew arrives at station 12 for rake/soil and documented samples, photo panorama	150:52		22/7:46 pm
Crew drives to station 13 (outer ejecta blanket, North Ray Crater)	151:47		22/8:41 pm
Crew arrives at station 13 for rock/soil samples, photo panorama	151:52		22/8:46 pm
Crew drives to station 14 (Smoky Mountain)	152:02		22/8:56 pm
Crew arrives at station 14 for rake/soil, double core, documented samples, photo panorama	152:09		22/9:03 pm
Crew drives to station 15 (Cayley Plains)	152:49		22/9:43 pm
Crew arrives at station 15 for magnetometer measurements, rock/soil sample, photo panorama	153:00		22/9:49 pm
Crew drives to station 16 (Dot Crater)	153:10		22/10:04 pm
Crew arrives at station 16 for magnetometer measurement, rock/soil samples, photo panorama	153:19		22/10:13 pm

Events	GET hrs:min	Date/EST
Crew drives to station 17 (NE rim, Palmetto Crater)	153:29	Apr 22/10:23 pm
Crew arrives at station 17 for rake/soil and documented samples, magnetometer readings, photo panorama	153:34	22/10:28 pm
Crew drives back to LM	154:12	22/11:06 pm
Crew arrives at LM	154:30	22/11:24 pm
Crew transfers samples, film packs, etc. stowage in LM; close-out final EVA	154:36	22/11:30 pm
LMP ingresses LM	155:19	23/00:13 am
CDR ingresses LM	155:23	23/00:17 am
Repressurize LM, end EVA 3	155:25	23/00:19 am

- more -

APOLLO 16
WALKING TRAVERSE
(CONTINGENCY)

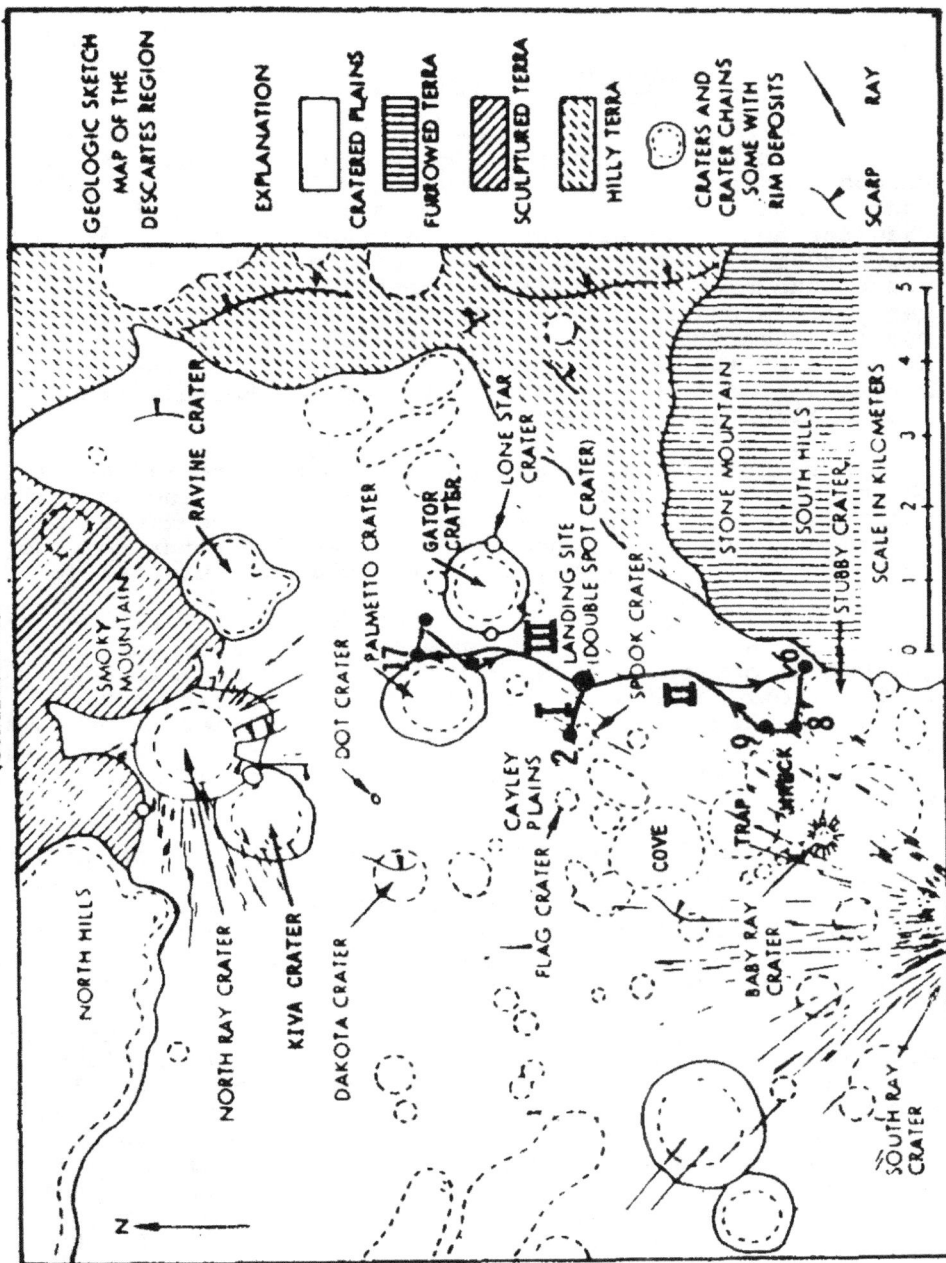

GEOLOGIC SKETCH MAP OF THE DESCARTES REGION

EXPLANATION

CRATERED PLAINS

FURROWED TERRA

SCULPTURED TERRA

HILLY TERRA

CRATERS AND CRATER CHAINS SOME WITH RIM DEPOSITS

SCARP   RAY

APOLLO 16 MISSION OBJECTIVES

Apollo 16 astronauts will explore the Descartes region, man's first opportunity to explore the lunar highlands. The site is some 2,250 meters (7,400 feet) higher than the Apollo 11 site and is representative of over three-fourths of the lunar surface. Preliminary geological analysis of the highlands indicates that the Moon's crust underwent modification early in its history. By studying these modification processes, we hope to achieve a better understanding of the development of this portion of the Moon's surface as well as the development of the Earth's crust, its continents, and ocean basins.

The three basic objectives are to explore and sample the materials and surface features, to set up and activate experiments on the lunar surface which will continue to relay data back to Earth after the crew has returned, and to conduct inflight experiments and photographic tasks.

The lunar roving vehicle, used for the first time on Apollo 15, will extend the range of the exploration and geological investigations that Young and Duke will make during their three seven-hour EVAs. The Apollo lunar surface package (ALSEP) which the crew will set up and place into operation will, with the Apollo 12, 14 and 15 ALSEPs, become the fourth in a network of lunar surface scientific stations.

The scientific instrument module (SIM) bay in the service module is the heart of the inflight experiment effort on Apollo 16. Quite similar to the SIM bay flown on Apollo 15, the bay contains high-resolution and mapping cameras and scientific sensors for photographing and measuring properties of the lunar surface and the environment around the Moon.

While in lunar orbit, command module pilot Mattingly will have the responsibility for operating the inflight experiments during the time his crewmates are on the lunar surface. During the homeward coast after transearth injection, Mattingly will exit through the command module crew hatch and maneuver hand-over-hand to the SIM bay where he will retrieve film cassettes, pass them back into the cabin for return to Earth and emplace a medical experiment for exposure to the solar-space environemnt. Mattingly also will take photographs aimed at gathering knowledge of astronomical phenomena such as Gegenschein from lunar orbit and ultraviolet photography of both the Earth and the Moon.

- more -

Engineering and operational tasks the Apollo 16 crew will carry out include further evaluation of the lunar roving vehicle and Skylab crew equipment, and use of the SIM bay subsatellite as a navigation tracking aid. Other medical experiments include biostack, ALFMED, and the passive bone mineral measurement.

SCIENTIFIC RESULTS OF APOLLO 11, 12, 14 and 15 MISSIONS

The Apollo 14 and Apollo 15 missions have clearly demonstrated that the major scientific returns from the Apollo program are to be expected from the later Apollo missions. The scientific results from the Apollo 15 mission have advanced the science of understanding the Moon from the stage characterized by interesting and stimulating speculations to one characterized by scientific hypotheses. The hypotheses which lead to specific experiments that, in turn, give answers that rapidly reduce the uncertainty regarding many basic questions concerning the origin and evolution of the Moon. The extended stay times and extended mobility available on the lunar surface and the wide-array of instrumentation carried in the SIM bay of the command and service module have already provided a rich harvest of scientific information from the first J mission.

Samples from the first two lunar landings showed us that most of the surface of the Moon dates back to a time when the terrestrial geologic record is just barely decipherable. They show that the lunar maria that covers about 1/3 of the side of the Moon viewed from Earth largely consists of an iron-rich volcanic rock produced from a partially molten shell 100-200 miles beneath the surface of the Moon. The soil samples from these two sites also included a variety of intriguing fragments clearly distinguishable from the mare basalts. These soil samples became the basis of widespread speculation on the composition and origin of the lunar highlands that make up most of the backside and approximately 2/3 of the front side of the Moon. Both the samples and the results of experiments flown in the Apollo 15 CSM have shown that these intriguing bits of rock are indeed representative of the large areas of the Moon that contain the bulk of the early history of the planet.

The measurement of gamma rays, produced by minute amounts of radioactive substances in the lunar soil, and characteristic x-rays, induced by the high energy solar radiation impinging on the sunlit side of the Moon has shown that much of the northwestern quadrant of the Earth side of the Moon is probably underlain by a very old (probably first formed 4.4-4.5 billion years ago) uranium- and thorium-rich volcanic rock and that most of the backside of the Moon, along with the eastern highland regions on the front side, appears to be made up of an aluminum- and calcium-rich rock (anorthosite) that was first observed in the Apollo 11 soil samples.

- more -

The widespread occurrence of both of these rock types
tells us a great deal about the early history of the Moon.
The aluminum-rich rocks suggest that the primitive Moon
had a liquid outer shell that may have been 50-80 miles
thick that gave rise to a lunar crust almost simultaneously
with the formation of the Moon itself. The old uranium-rich
volcanic rocks both limit the thickness of this early
liquid layer and tell us something about the composition
of the solid interior very early during lunar history.
In particular, they suggest that the Moon was formed from
material that condensed at temperatures much higher than
previously thought out of the primitive dust cloud that
surrounded the early Sun.

In addition to the chemical study of the lunar surface,
the Apollo 15 mission returned excellent photographs of more
than 10 percent of the surface of the Moon which reveal
geological features as small as the size of the LM. Like
the rock samples returned from this and previous missions,
these photographs will provide a source of data for scientists
for generations to come.

A subsatellite left behind by the Apollo 15 mission
has measured the present magnetic field over a large per-
centage of the Moon's surface. These magnetic field
measurements, along with studies of the magnetization of
lunar rocks, indicate that the Moon's magnetic field 3 billion years ago must have been 100 to 1,000 times stronger
than it is today. The subsatellite and altimeter have also
shown that the mascons discovered by Lunar Orbiter spacecraft consist of circular plates of very dense rocks which
fill the deep circular basins produced by collisions of
asteroid-size objects with the Moon.

Most of the experiments deployed on the lunar surface
during the Apollo 12, 14, and 15 missions continue to function well, but more important -- they have provided information regarding geological processes that are still going
on today and have shown us that the interior of the Moon can
be investigated with present instrumentation. The impact
of the Apollo 15 SIVB was recorded by both the Apollo 14 and
Apollo 12 seismometers. Sound waves from this impact pene-
trated about 80 kilometers (50 miles) into the lunar
interior. The relatively high sound velocity found at these
depths was a complete surprise to seismologists and lunar
scientists. The sound velocity profile in the upper 80
kilometers (50 miles) indicates that the concept of the lunar
crust derived from mineralogy and chemistry of rocks was well-
founded. Ultrasensitive thermometers implaced in two holes
on the lunar surface showed that the amount of heat escaping
from the lunar interior exceeds even the highest estimates made
by scientists who speculated on the content of radioactive
elements found in the Moon.

- more -

During the week of January 10, approximately 700 scientists, 15 from foreign countries including three scientists from the Soviet Union, gathered at the Manned Spacecraft Center to present and summarize results obtained by the Apollo 14, Apollo 15, and Luna 16 missions. In addition to the conclusions and ideas already mentioned here, studies of Apollo 14 samples have shown that the largest collision with an asteroidal object recorded on the surface of the Moon took place about 4 billion years ago. This was much later than had been anticipated from various theories dealing with the formation of large planetary objects. It raises the possibility that major collisions of asteroid-sized objects and the Earth continued more than 1 billion years after the planet was formed.

The relatively small Luna 16 samples provided a surprising number of extremely interesting results that indicate that unmanned sampling of the Moon is, indeed, a useful augmentation of the manned program.

As we enter into the homestretch of the Apollo program, it is clear that a successful Apollo 16 and 17 added to the missions already flown will produce a chapter in the history of science that will be a permanent and living testimony to this undertaking.

- more -

APOLLO 16 LANDING SITE

A hilly region north of the Descartes crater in a highlands area of the southeastern quadrant of the visible face of the Moon is the landing site chosen for Apollo 16.

The Descartes site appears to have structural characteristics similar to vulcanism sites on Earth, and has two separate volcanic features -- Cayley Plains and the Descartes mountains -- which will be extensively explored and sampled by the Apollo 16 crew.

The Cayley Plains segment of the landing site is characterized by terrain ranging from smooth to undulating, -- possibly as a result of fluid volcanic rock flow. The Descartes Mountains, part of the Kant Plateau are characterized by hilly, furrowed highland plateau material that is thought to have come from a more viscous volcanic flow. Additionally, the Descartes landing site provides an opportunity to study the evolution of young, bright-rayed craters and to extend age-dating to similar craters in other regions of the Moon.

The landing site has two basic areas which will be explored and sampled: Cayley Plains, including North Ray and South Ray craters; Stone Mountain and Smoky Mountain of the South and North Descartes Mountains.

The low crater density in the Cayley Plains suggests an Imbrian age for the rolling, ridged portion of Cayley in the Apollo 16 traverse area. Stone and Smoky mountains, on the other hand, appear to have shapes typical of volcanic formations on Earth -- shapes that might be formed by movement of rather viscous material.

North and South Ray craters appear to penetrate deeply into the Cayley formation and reveal the sequence of layering, perhaps through an overlap of both the Cayley and Descartes formations.

Smaller, subdued craters in the landing site seem to have a characteristic concave bottom, which suggests that the substrata underlying the crater impact were more resistant than in other crater fields on the Moon.

- more -

APOLLO 16 TRAVERSES – GEOLOGIC SKETCH

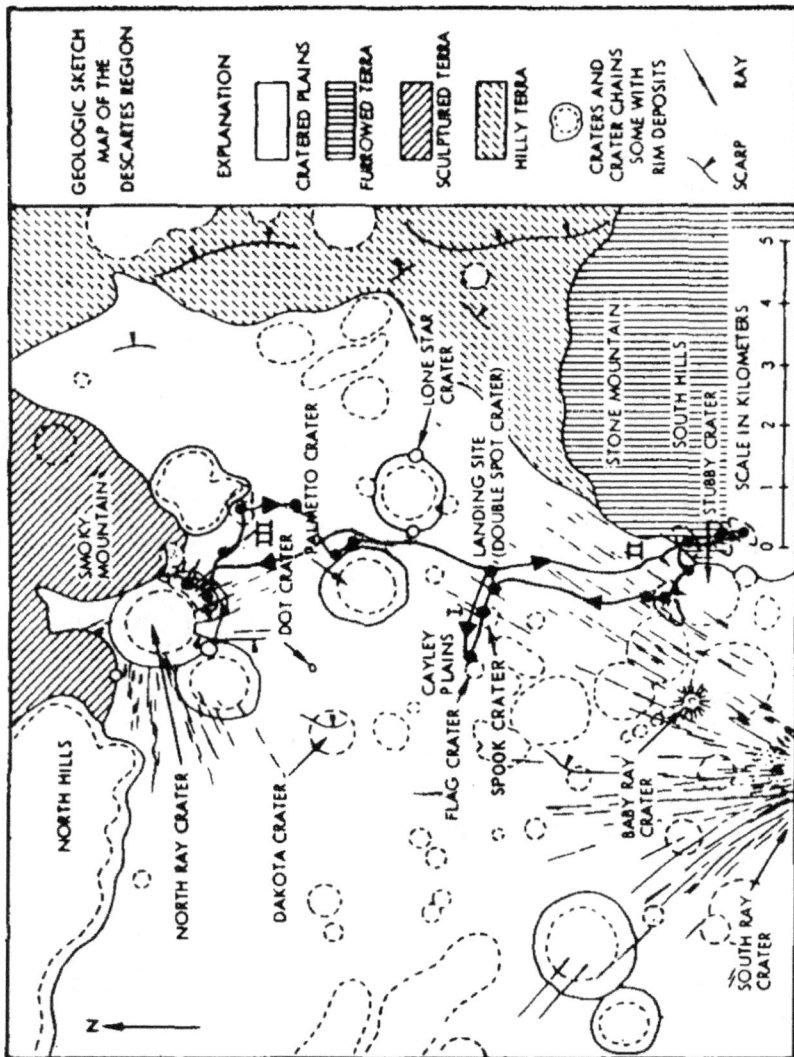

GEOLOGIC SKETCH MAP OF THE DESCARTES REGION

EXPLANATION

CRATERED PLAINS

FURROWED TERRA

SCULPTURED TERRA

HILLY TERRA

CRATERS AND CRATER CHAINS SOME WITH RIM DEPOSITS

SCARP

RAY

NORTH HILLS

NORTH RAY CRATER

DAKOTA CRATER

SMOKY MOUNTAIN

DOT CRATER

PALMETTO CRATER

LONE STAR CRATER

LANDING SITE (DOUBLE SPOT CRATER)

CAYLEY PLAINS

FLAG CRATER

SPOOK CRATER

BABY RAY CRATER

SOUTH RAY CRATER

STONE MOUNTAIN

SOUTH HILLS

STUBBY CRATER

SCALE IN KILOMETERS

0 1 2 3 4 5

SITE	LAT.	LONG.	TERRAIN FEATURES	SCIENTIFIC INTEREST
APOLLO 11 (SEA OF TRANQUILITY)	0°41'N.	23°26'E.	MARE: LARGE SUBDUED CRATERS (200-600 METERS DIAMETER), INTERMEDIATE CRATERS (50-200 METERS DIAMETER).	MARE MATERIAL AGE AND COMPOSITION, GEOPHYSICAL DATA.
APOLLO 12 (OCEAN OF STORMS)	3°12'S.	23°24'W	MARE: SAME AS APOLLO 11.	MATERIAL AGE AND COMPOSITION FOR COMPARISON WITH APOLLO 11, SURVEYOR DATA.
FRA MAURO (APOLLO 14)	3°40'S.	17°28'W	UPLAND: DEEP-SEATED EJECTA BLANKET.	EJECTA AGE, COMPOSITION AND STRUCTURE, GEOPHYSICAL DATA.
HADLEY-APENNINES (APOLLO 15)	26°04'N.	3°39'E	MARE/HIGHLAND: V-SHAPED SINUOUS RILLE, RILLE FLOOR BLOCKS; APENNINE HIGHLANDS.	BASIN AND RILLE ORIGIN AND AGE; HIGHLANDS COMPOSITION, GEOPHYSICAL DATA
DESCARTES	10°00'S.	16°00'E	HIGHLAND/HIGHLAND BASIN FILL	VOLCANICS AND PLAINS STRUCTURE, HIGHLAND AGE AND COMPOSITION, VOLCANISM TIME SPAN AND COMPOSITIONAL TRENDS, GEOPHYSICAL DATA.

APOLLO LANDING SITES

Lunar Surface Science

The ALSEP array carried on Apollo 16 has four experiments: heat flow, lunar surface magnetometer, passive seismic and active seismic.

Six additional experiments will be conducted on the Descartes landing area: far ultraviolet camera/spectroscope, portable magnetometer, cosmic ray detector (sheets), solar wind composition, lunar geology investigation, and soil mechanics.

Passive Seismic Experiment (PSE): The PSE measures seismic activity of the Moon and gathers and relays to Earth information relating to the physical properties of the lunar crust and interior. The PSE reports seismic data on man-made impacts (LM ascent stage), natural impacts of meteorites, and moon-quakes. Dr. Gary Latham of the Lamont-Doherty Geological Observatory (Columbia University) is responsible for PSE design and experiment data analysis.

Three similar PSEs, deployed as a part of the Apollo 12, 14 and 15 ALSEPs, have transmitted to Earth data on lunar surface seismic events since deployment. The Apollo 12, 14, 15 and 16 seismometers differ from the seismometer left at Tranquillity Base in July 1969 by the Apollo 11 crew in that the later PSEs are continuously powered by SNAP-27 radio-isotope thermoelectric generators. The Apollo 11 seismometer, powered by solar cells, transmitted data only during the lunar day and is no longer functioning.

After Apollo 16 trans-lunar injection, the Saturn V's spent S-IVB stage and instrument unit will be aimed to impact the Moon. This will stimulate the passive seismometers left on the lunar surface during previous Apollo missions.

The S-IVB/IU will be commanded to hit the Moon 250 kilo-meters (135 nautical miles) west of the Apollo 12 ALSEP site, at a target point 2.2 degrees south latitude by 31.4 degrees west longitude, near Lansberg Craters F and D in the Ocean of Storms.

After the spacecraft is ejected from the launch vehicle, a launch vehicle auxiliary propulsion system (APS) ullage motor will be fired to separate the vehicle a safe distance from the spacecraft. Residual liquid oxygen in the almost spent S-IVB/IU will then be dumped through the engine with the vehicle positioned so the dump will slow it into an impact trajectory. Mid-course corrections will be made with the stage's APS ullage motors if necessary.

	APOLLO MISSIONS					
	11	12	14	15	16	17
ALSEP EXPERIMENTS						
S-031 LUNAR PASSIVE SEISMOLOGY	X	X	X	X	X	
S-033 LUNAR ACTIVE SEISMOLOGY			X		X	
S-034 LUNAR TRI-AXIS MAGNETOMETER		X		X	X	
S-035 SOLAR WIND SPECTROMETER		X		X		
S-036 SUPRATHERMAL ION DETECTOR		X	X	X		
S-037 LUNAR HEAT FLOW				X	X	X
S-038 CHARGED PARTICLE LUNAR ENVIRONMENT			X			
S-058 COLD CATHODE GUAGE		X	X	X		
S-202 LUNAR EJECTA AND METEORITES						X
S-203 LUNAR SEISMIC PROFILING						X
S-205 LUNAR ATMOSPHERIC COMPOSITION						X
S-207 LUNAR SURFACE GRAVIMETER						X
M-515 LUNAR DUST DETECTOR	X	X	X	X		
OTHER EXPERIMENTS						
S-059 LUNAR GEOLOGY INVESTIGATION	X	X	X	X	X	X
S-078 LASER RANGING RETRO-REFLECTOR	X		X	X		
S-080 SOLAR WIND COMPOSITION	X	X	X	X	X	
S-151 COSMIC RAY DETECTOR (HELMET)	X					
S-152 COSMIC RAY DETECTOR (SHEETS)					X	
S-184 LUNAR SURFACE CLOSE-UP CAMERA	X	X	F			
S-198 PORTABLE MAGNETOMETER			X		X	
S-199 LUNAR GRAVITY TRAVERSE						X
S-200 SOIL MECHANICS		X	X	X	X	X
S-201 FAR UV CAMERA/SPECTROSCOPE					X	X
S-204 SURFACE ELECTRICAL PROPERTIES						X

F - FACILITY EQUIPMENT

LUNAR SURFACE SCIENCE EXPERIMENT ASSIGNMENTS

Lunar Surface Experiment Scientific Discipline
Contribution Correlation

SCIENTIFIC DISCIPLINE / EXPERIMENT	GEOLOGY	GEOPHYSICS	GEOCHEMISTRY	BIOSCIENCES	LUNAR ATMOSPHERE	PARTICLES AND FIELDS	ASTRONOMY
CONTINGENCY SAMPLE COLLECTION	AID IN DETERMINING LUNAR HISTORY BY AGING OF LUNAR SAMPLES		DETERMINE COMPOSITION OF LUNAR SURFACE BY CHEMICAL ANALYSIS OF LUNAR SAMPLES	AID IN DETERMINING POSSIBILITY OF BIOLOGICAL FORMS ON LUNAR SURFACE		MONOPOLE SEARCH	
ALSEP PASSIVE SEISMIC (S-031)	AID IN DETERMINING INTERIOR STRUCTURE, TECTONISM AND VOLCANISM	AID IN DETERMINING FREE OSCILLATIONS, TIDES, SECULAR STRAINS, TILT, VELOCITY, ATTENUATION AND DIRECTION OF SEISMIC WAVES					MEASURE METEOROID IMPACTS
ACTIVE SEISMIC (S-033)	AID IN DETERMINING THE TYPE AND CHARACTER AS WELL AS HARDNESS AND BEARING STRENGTH OF LUNAR MATERIALS	AID IN DETERMINING FREE OSCILLATIONS, TIDES, SECULAR STRAINS, TILT, VELOCITY, ATTENUATION AND DIRECTION OF SEISMIC WAVES					MEASURE METEOROID IMPACTS
LUNAR SURFACE MAGNETOMETER (S-034)	AID IN DETERMINING MAGNETIC ANOMALIES, SUBSURFACE FEATURES AND LUNAR HISTORY	AID IN DETERMINING THERMAL STATE OF THE LUNAR INTERIOR				ESTABLISH GROSS ELECTRICAL DIFFUSIVITY, MEASURE MAGNETIC FIELD OF THE MOON	DETERMINE LUNAR RESPONSE TO FLUCTUATIONS IN THE INTERPLANETARY MAGNETIC FIELD
HEAT FLOW (S-037)	AID IN DETERMINING LUNAR EVOLUTION	MEASURE VERTICAL TEMPERATURE GRADIENTS, ABSOLUTE TEMPERATURE OF THE SURFACE TO ESTABLISH VERTICAL THERMAL CONDUCTIVITY	BULK COMPOSITION AND CHEMICAL SORTING MAY BE INFERRED FROM DATA				DETERMINE THERMAL ENVIRONMENT
PORTABLE MAGNETOMETER (S-198)	AID IN DETERMINING LOCAL LUNAR MAGNETIC ANOMALIES, PRESENCE OF MASCONS AND OTHER LOCAL MAGNETIC FIELD SOURCES	AID IN MEASURING DISTORTION OF EARTH MAGNETIC FIELD BY SOLAR WIND, SHOCK FRONT, BOUNDARY LAYERS, MAGNETOPAUSE.				MEASURE LOCAL MAGNETIC FIELD OF MOON AND MAGNETIC FIELD/SOLAR PLASMA INTERACTION.	DETERMINE LUNAR FIELD FLUCTUATIONS DUE TO DIFFUSION OF INTERPLANETARY MAGNETIC FIELD.
FAR UV CAMERA/SPECTROSCOPE		PROVIDE INFORMATION ON THE DENSITY, DISTRIBUTION, AND COMPOSITION OF GEO-CORONA, SOLAR WIND, AND INTERSTELLAR MEDIUM			DETECT THE POSSIBLE PRESENCE OF A LUNAR ATMOSPHERE OR GASS VENTING FROM THE LUNAR SURFACE		PROVIDE OPPORTUNITY TO DETECT RED-SHIFTED EMISSION FROM INTERGALACTIC HYDROGEN, AND OBTAIN AN ACCURATE SURVEY OF FAINT SOURCES OF COSMIC ULTRAVIOLET LIGHT
COSMIC RAY DETECTOR (SHEETS) (S-152)		AID IN DETERMINING THE ORIGIN OF TEKTITES (GLASSY BODY OF PROBABLY METEORITIC ORIGIN)	DETERMINE COMPOSITION OF SOLAR WIND PLASMA		AID IN DETERMINING HISTORY OF PLANETARY ATMOSPHERE	MEASURE NEUTRON FLUX ON LUNAR SURFACE, PROVIDE DATA ON COSMIC RAY PARTICLES, BOTH SOLAR AND GALACTIC.	
LUNAR GEOLOGY INVESTIGATION S-059	AID IN DETERMINING LUNAR GEOLOGICAL STRUCTURE AND HISTORY		DETERMINE CHEMICAL COMPOSITION OF LUNAR SAMPLES	LUNAR SAMPLES MAY BE TESTED FOR ABILITY TO SUPPORT LIFE FORMS USED TO DETERMINE POSSIBILITY OF BIOLOGICAL LIFE FORMS ON THE LUNAR SURFACE		MONOPOLE SEARCH	
SOIL MECHANICS S-200	AID IN DETERMINING LUNAR HISTORY, ENABLE DETERMINATION OF COMPOSITIONAL TEXTURAL, AND MECHANICAL PROPERTIES OF LUNAR SOIL		ENABLE DETERMINATION OF COMPOSITION OF LUNAR SOIL				
SOLAR WIND COMPOSITION (S-080)			DETERMINE COMPOSITION OF SOLAR WIND PLASMA		AID IN DETERMINING HISTORY OF PLANETARY ATMOSPHERE	PROVIDE INFORMATION ON THE ELEMENTAL AND ISOTOPIC COMPOSITION OF NOBLE GASES AND OTHER ELEMENTS IN THE SOLAR WIND	

APOLLO 16 ALSEP DEPLOYMENT

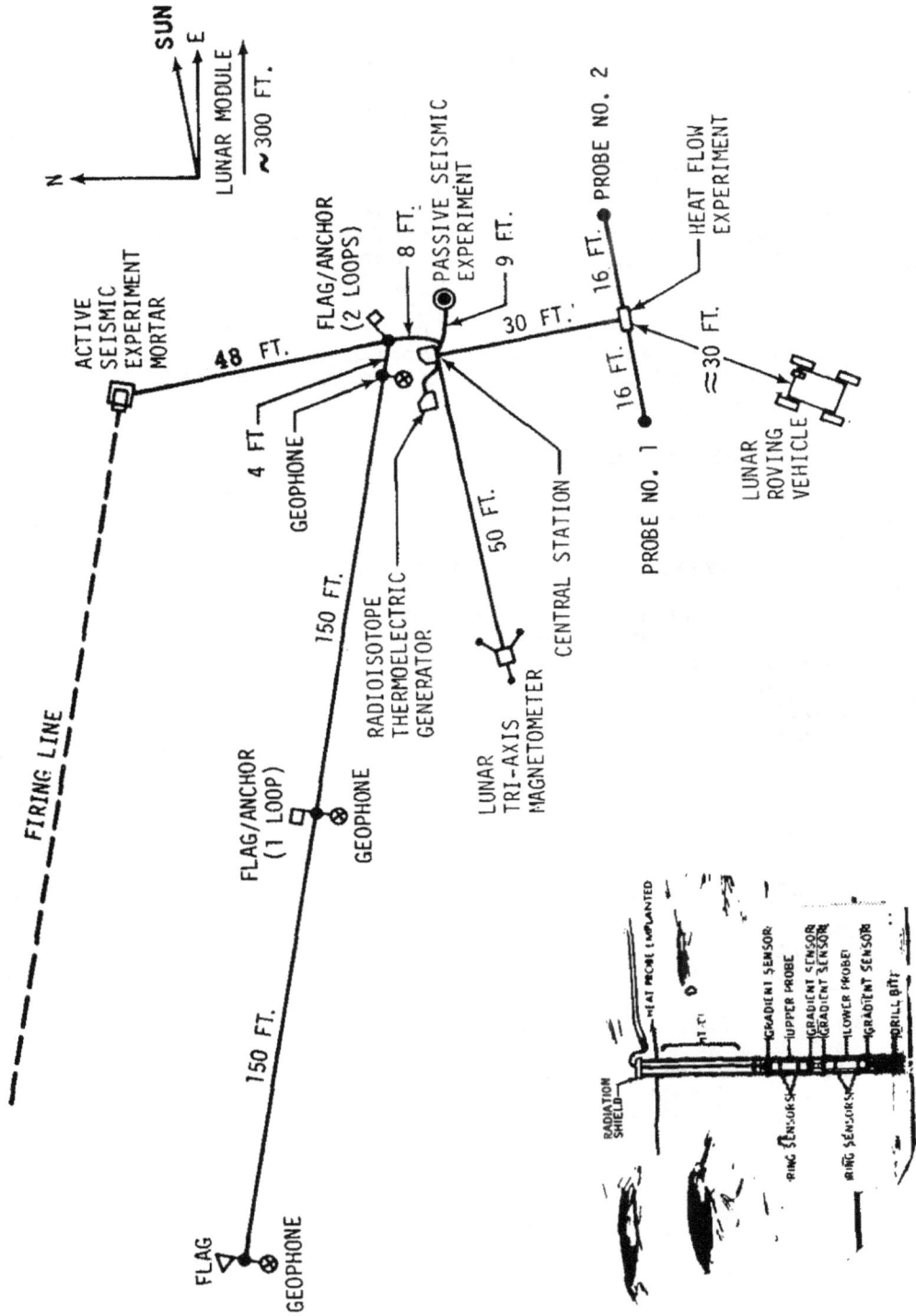

The S-IVB/IU will weigh 13,973 kilograms (30,805 pounds) and will be traveling 9,223 kilometers an hour (4,980 nautical mph) at lunar impact. It will provide an energy source at impact equivalent to about 11 tons of TNT.

After Young and Duke have completed their lunar surface operations and rendezvoused with the command module in lunar orbit, the lunar module ascent stage will be jettisoned and later ground-commanded to impact on the lunar surface west of the Apollo 16 landing site at Descartes. The stage will impact the surface at 1.692 m/s (5,550 fps) at a -3.2° angle.

Impacts of these objects of known masses and velocities will assist in calibrating the Apollo 16 PSE readouts as well as in providing comparative readings between the Apollo 12, 14, 15 and 16 seismometers.

ALSEP to Impact Distance Table

	Approximate Distance	
	Kilometers	Nautical miles
Apollo 12 ALSEP (3.03°S, 23.4°W) to:		
Apollo 12 LM A/S Impact	72	39
Apollo 13 S-IVB Impact	137	74
Apollo 14 S-IVB Impact	175	95
Apollo 14 LM A/S Impact	114	62
Apollo 15 S-IVB Impact	354	191
Apollo 15 LM A/S Impact	1130	610
Apollo 16 S-IVB Impact	252	136
Apollo 16 LM A/S Impact	1173	633
Apollo 14 ALSEP (3.67°S, 17.45°W) to:		
Apollo 14 LM A/S Impact	68	37
Apollo 15 S-IVB Impact	184	99
Apollo 15 LM A/S Impact	1048	565
Apollo 16 S-IVB Impact	433	234
Apollo 16 LM A/S Impact	992	536
Apollo 15 ALSEP (26.10°N, 3.65°E) to:		
Apollo 15 LM A/S Impact	92	50·
Apollo 16 S-IVB Impact	1347	727
Apollo 16 LM A/S Impact	1129	609

-more-

S-IVB/IU IMPACT

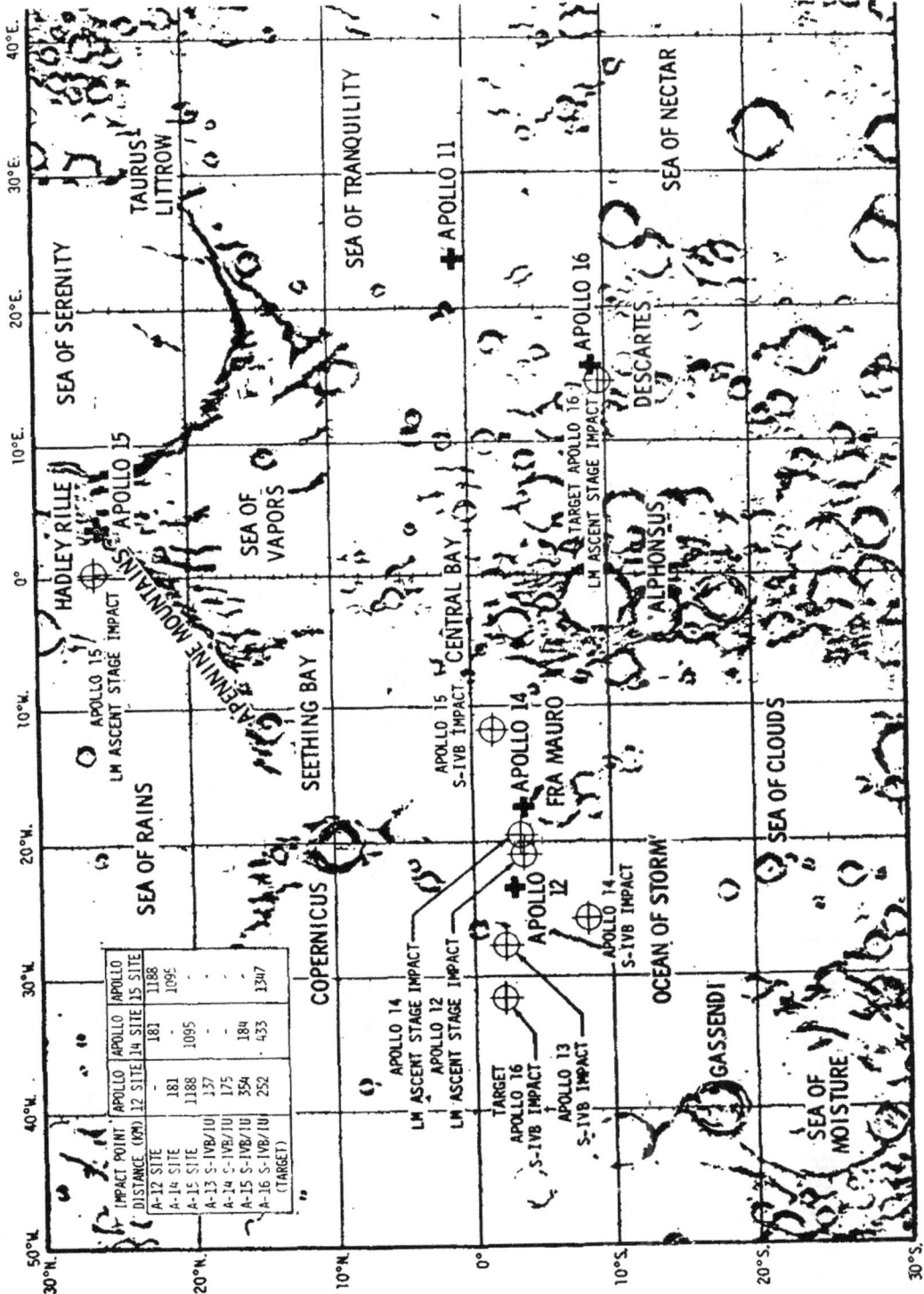

IMPACT POINT DISTANCE (KM)	APOLLO 12 SITE	APOLLO 14 SITE	APOLLO 15 SITE
A-12 SITE	-	181	1188
A-14 SITE	181	-	1095
A-15 SITE	1188	1095	-
A-13 S-IVB/IU	137	-	-
A-14 S-IVB/IU	175	184	-
A-15 S-IVB/IU	354	-	-
A-16 S-IVB/IU (TARGET)	252	433	1347

LUNAR IMPACT TARGET FOR SPENT LM ASCENT STAGE

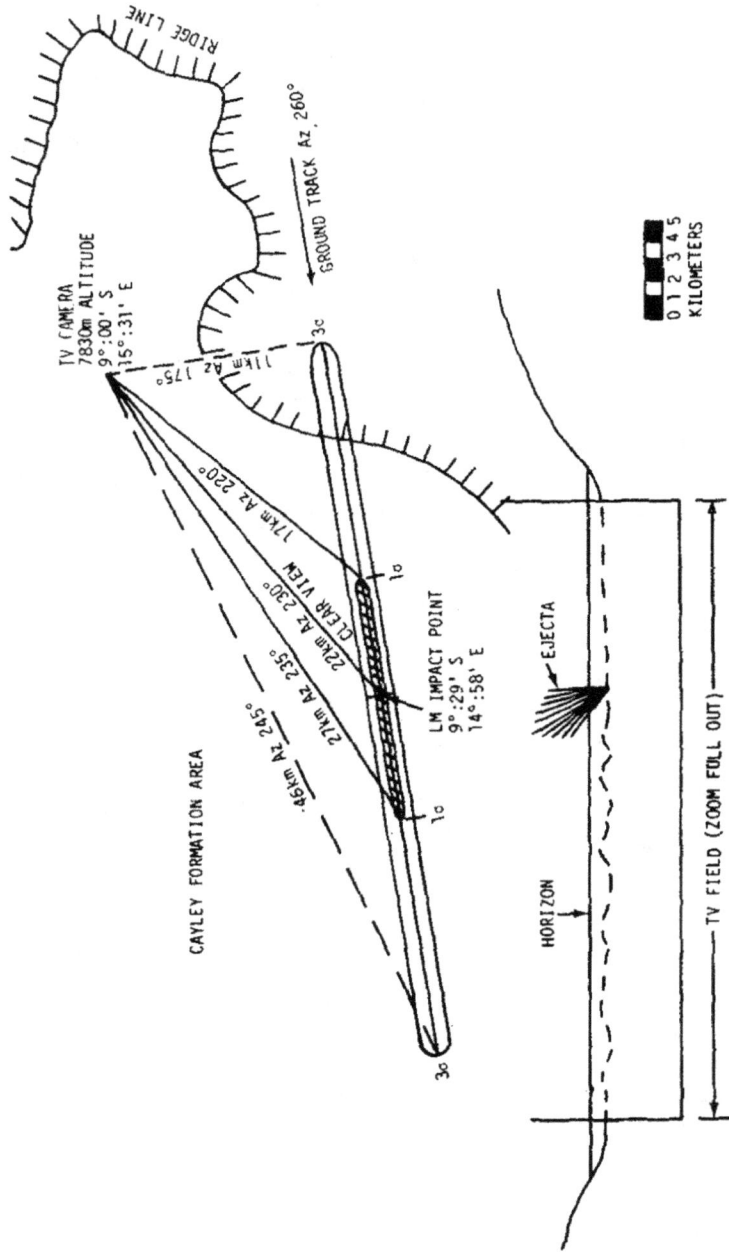

RIDGE LINE

GROUND TRACK AZ 260°

TV CAMERA
7830m ALTITUDE
9°:00' S
15°:31' E

11km AZ 175°

17km AZ 220°

CLEAR VIEW

20km AZ 230°

27km AZ 235°

46km AZ 245°

CAYLEY FORMATION AREA

LM IMPACT POINT
9°:29' S
14°:58' E

EJECTA

HORIZON

TV FIELD (ZOOM FULL OUT)

0 1 2 3 4 5
KILOMETERS

Approximate Distance

Kilometers   Nautical Miles

Apollo 16 ALSEP (9.0^{O}S, 15.52^{O}E) to:

    Apollo 16 LM A/S Impact   23   12
    Apollo 16 S-IVB Impact 2.3^{O}S, 31°W
    Apollo 16 LM A/S Impact 9.48^{O}S, 14.97^{O}E

There are three major physical components of the PSE:

1.   The sensor assembly is conprised of 3 long period (LP)
and a short period (SP) seismometer, an electrical power and
a data subsystem, and a thermal control system. In the LP
seismometer, low frequency (approximately 250 to 0.3 second
periods) motion of the lunar surface caused by seismic activity is detected by tri-axial, orthogonal displacement ampli-
tude type sensors. In the SP seismometer, the higher frequency (approximately 5 to 0.04 second periods) vertical motion
of the lunar surface is detected by a displacement velocity
sensor.

2.   The external leveling stool allows manual leveling of
the sensor assembly by the crewman to within +5 degrees.
Final internal leveling to within +3 arc seconds is accomplished by control motors.

3.   The five-foot diameter hat-shaped thermal shroud covers
and helps stabilize the temperature of the sensor assembly.
The instrument uses thermostatically controlled heaters to
protect it from the extreme cold of the lunar night.

The Lunar Surface Magnetometer (LSM): The scientific objective of the magnetometer experiment is to measure the magnetic
field at the lunar surface. Charged particles and the magnetic field of the solar wind impact directly on the lunar
surface. Some of the solar wind particles are absorbed by
the surface layer of the Moon. Others may be deflected a-
round the Moon. The electrical properties of the material
making up the Moon determine what happens to the magnetic
field when it hits the Moon. If the Moon is a perfect insul-
ator the magnetic field will pass through the Moon undisturbed.
If there is material present which acts as a conductor,
electric currents will flow in the Moon. A small magnetic
field of approximately 35 gammas , one thousandth the size
of the Earth's field was recorded at the Apollo 12 site.

Fields recorded by the portable magnetometer on Apollo 14
were about 43 gammas and 103 gammas in two different locations. (Gamma is a unit of intensity of a magnetic field.
The Earth's magnetic field at the Equator, for example, is
35,000 gamma. The interplanetary magnetic field from the
Sun has been recorded at 5 to 10 gamma.)

Two possible models are shown in the next drawing. The
electric current carried by the solar wind goes through the
Moon and "closes" in the space surrounding the Moon (figure a).
This current (E) generates a magnetic field (M) as shown.
The magnetic field carried in the solar wind will set up a
system of electric currents in the Moon or along the surface.
These currents will generate another magnetic field which
tries to counteract the solar wind field (figure b). This
results in a change in the total magnetic field measured at
the lunar surface.

The magnitude of this difference can be determined by
independently measuring the magnetic field in the undisturbed
solar wind nearby, yet away from the Moon's surface. The
value of the magnetic field change at the Moon's surface can
be used to deduce information on the electrical properties
of the Moon. This, in turn, can be used to better understand
the internal temperature of the Moon and contribute to better
understanding of the origin and history of the Moon.

The principal invesitgator for this experiment is
Dr. Palmer Dyal, NASA Ames Research Center, Mountain View,
California.

The magnetometer consists of three magnetic sensors
aligned in three orthogonal sensing axes, each located at
the end of a fiberglass support arm extending from a central
structure. This structure houses both the experiment electronics and the electro-mechanical gimbal/flip unit which
allows the sensor to be pointed in any direction for site
survey and calibration modes. The astronaut aligns the magnetometer experiment to within +3 degrees east-west using a
shadowgraph on the central structure, and to within +3 degrees
of the vertical using a bubble level mounted on the $\overline{Y}$ sensor
boom arm

LUNAR MAGNETIC ENVIRONMENT

MAGNETIC FIELD OF MOON
(M) GENERATED BY THE ELECTRIC
FIELD (E) CARRIED IN THE SOLAR WIND

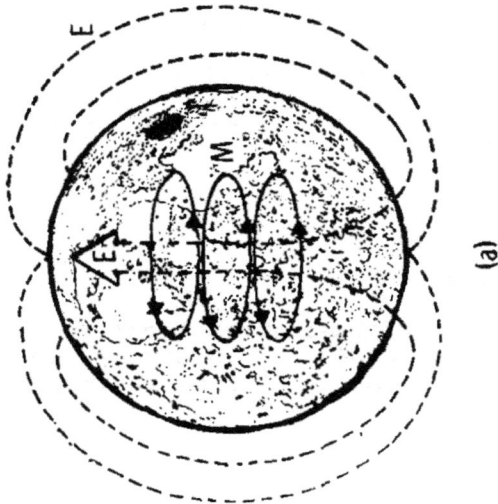

(a)

MAGNETIC FLUX CARRIED IN
THE SOLAR WIND INDUCES
EDDY CURRENTS (E) WHICH
IN TURN INDUCES A MAGNETIC
FIELD

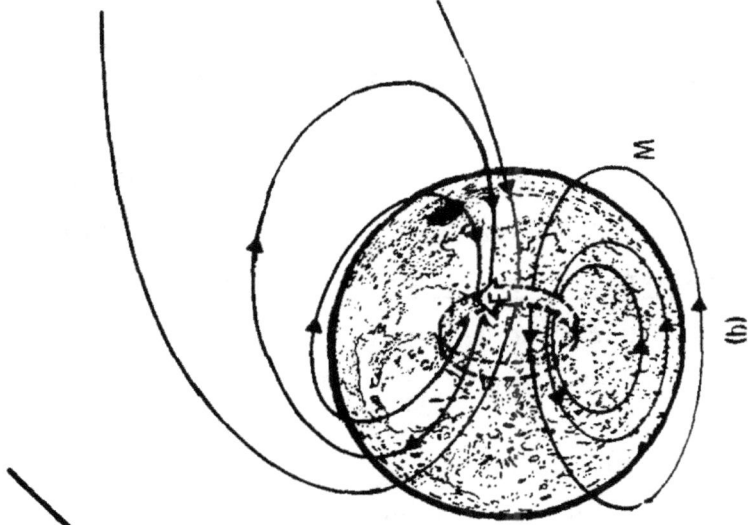

(b)

Size, weight and power are as follows:

Size deployed   102 cm (40 inches) high
                                     with 152 cm (60 inches)
                                     between sensor heads

Weight (pounds)   8 kilograms (17.5 lbs)

Peak Power Requirements   (watts)

Site Survey Mode   11.5

Scientific Mode   6.2
                                     12.3   (night)

Calibration Mode   10.8

The magnetometer experiment operates in three modes:

Site Survey Mode -- An initial site survey is performed
in each of the three sensing modes for the purpose of locating
and identifying any magnetic influences permanently inherent
in the deployment site so that they will not affect the interpretation of the LSM sensing of magnetic flux at the lunar
surface.

Scientific Mode -- This is the normal operating mode
wherein the strength and direction of the lunar magnetic field
are measured continously. The three magnetic sensors provide
signal outputs proportional to the incidence of magnetic field
components parallel to their respective axes. Each sensor
will record the intensity three times per second which is faster than the magnetic field is expected to change. All sensors
have the capability to sense over any one of three dynamic
ranges with a resolution of 0.2 gamma.

-100 to +100 gamma

-200 to +200 gamma

-400 to +400 gamma

Calibration Mode -- This is performed automatically at
18-hour intervals to determine the absolute accuracy of the
magnetometer sensors and to correct any drift from their laboratory calibration.

Magnetic Lunar Sample Returned to the Moon

In the study of returned lunar samples it has been found
that there are commonly two components of remanent magnetization. The first of these is quite soft and can be removed
by cleaning in alternating fields of about 50 to 100 oersteds.
The second component is quite stable and is hardly affected
by alternating fields up to 500 oersteds. It is this stable
component which is most likely to represent ancient lunar
magnetic fields and on which most attention has been focused.
It can be simulated in the laboratory by allowing the sample
to cool in the presence of a weak field from above its Curie
point -- about 780 degrees Centigrade (1,400 degrees Fahrenheit)
in the case of lunar samples.

The soft component can also be simulated in the laboratory
by exposing lunar samples to steady magnetic fields of about
10 to 50 oersteds. The field required to do this is much
greater than the Earth's field which is only about 0.5 oersteds.
One possibility, therefore, is that this magnetization was
acquired on the return journey to Earth either in the spacecraft or in Earth laboratories. Another possibility is that
this magnetization is truly of lunar origin, perhaps related
to thermal fluctuations in the presence of a weak, solar wind
field. It is important to establish the origin of this soft
component and two steps are being taken to do this.

At the last lunar portable magnetometer (LPM) station a
documented sample (preferably igneous) will be picked up and
placed on the LPM after the first reading is taken. It will
be documented in place on the magnetometer and the LPM reading
retaken. If the intensity of magnetization of the soft component is like that seen in the returned samples it should be
measurable at the level of several gammas. On return to Earth
this same sample will be placed on the back-up flight model LPM
in the shielded room in the LRL and the same procedure repeated.
This will tell if additional magnetization has been acquired
on the return journey.

The second step consists of taking a lunar sample back to
the Moon. The sample chosen is Apollo 12 sample 12002,78.
Weight is 4 grams (0.14 ounces). This sample has been thoroughly
tested magnetically and has a very clear stable component of
magnetization. When first received in the laboratory it also
had a soft component of magnetization which has been removed.
Upon return to Earth it will be tested once again to see if it
has reacquired the soft component. If it has, the soft remanence
is not of lunar origin. The sample will be placed in a bag,
sewn inside an interim stowage assembly bag and returned to
Earth stowed as on Apollo 12.

- more -

Lunar Heat Flow Experiment (HFE): The scientific objective of the heat flow experiment is to measure the steady-state heat flow from the lunar interior. Two predicted sources of heat are: (1) original heat at the time of the Moon's formation and (2) radioactivity. Scientists believe that heat could have been generated by the infalling of material and its subsequent compaction as the Moon was formed. Moreover, varying amounts of the radioactive elements uranium, thorium and potassium were found in the Apollo 11, 12, 14, and 15 lunar samples which if present at depth, would supply significant amounts of heat. No simple way has been devised for relating the contribution of each of these sources to the present rate of heat flow. In addition to temperature, the experiment is capable of measuring the thermal conductivity of the lunar rock material.

The combined measurement of temperature and thermal conductivity gives the net heat flux from the lunar interior through the lunar surface. Similar measurements on Earth have contributed basic information to our understanding of volcanoes, earthquakes and mountain building processes. In conjuction with the seismic and magnetic data obtained on other lunar experiments the values derived from the heat flow measurements will help scientists to build more exact models of the Moon and thereby give us a better understanding of its origin and history.

The heat flow experiment consists of instrument probes, electronics and emplacement tool and the lunar surface drill. Each of two probes is connected by a cable to an electronics box which rests on the lunar surface. The electronics, which provide control, monitoring and data processing for the experiment, are connected to the ALSEP central station.

Each probe consists of two identical 51 cm (20 in.) long sections each of which contains a "gradient" sensor bridge, a "ring" sensor bridge and two heaters. Each bridge consists of four platinum resistors mounted in a thin-walled fiberglass cylindrical shell. Adjacent areas of the bridge are located in sensors at opposite ends of the 51 cm (20 in.) fiberglass probe sheath. Gradient bridges consequently measure the temperature difference between two sensor locations.

In thermal conductivity measurements at very low values a heater surrounding the gradient sensor is energized with 0.002 watts and the gradient sensor values monitored. The

-more-

rise in temperature of the gradient sensor is a function of thermal conductivity of the surrounding lunar material. For higher range of values, the heater is energized at 0.5 watts of heat and monitored by a ring sensor. The rate of temperature rise, monitored by the ring sensor is a function of the thermal conductivity of the surrounding lunar material. The ring sensor, approximately 10 cm (4 in.) from the heater, is also a platinum resistor. A total of eight thermal conductivity measurements can be made. The thermal conductivity mode of the experiment will be implemented about 20 days (500 hours) after deployment. This is to allow sufficient time for the perturbing effects of drilling and emplacing the probe in the borehole to decay; i.e., for the probe and casings to come to equilibrium with the lunar subsurface.

A 9.1 meter (30-ft.) cable connects each probe to the electronics box. The cable contains four evenly spaced thermocouples: at the top of the probe; at 66 cm (26 in.), 114 cm (45 in.) and 168 cm (66 in.). The thermocouples will measure temperature transients propagating downward from the lunar surface. The reference junction temperature for each thermocouple is located in the electronics box. In fact, the feasibility of making a heat flow measurement depends to a large degree on the low thermal conductivity of the lunar surface layer, the regolith. Measurement of lunar surface temperature variations by Earth-based telescopes as well as the *Surveyor* and *Apollo* missions show a remarkably rapid rate of cooling. The wide fluctuations in temperature of the lunar surface are to influence only the upper one meter (three ft.) and not the bottom one meter (three ft.) of the borehole.

The astronauts will use the Apollo lunar surface drill (ALSD) to make a lined borehole in the lunar surface for the probes. The drilling energy will be provided by a battery-powered rotary percussive power head. The drill borestem rod consists of fiberglass tubular sections reinforced with boron filaments, one section 137 cm (54 in) long and two sections each 71 cm (28 in) long. A closed drill bit, placed on the first drill rod, is capable of penetrating the variety of rock including 95 cm (3 ft.) of vesicular basalt (40 percent porosity). As lunar surface penetration progresses, additional drill rod sections will be connected to the drill string. The drill string is left in place to serve as a hole casing

An emplacement tool is used by the astronaut to insert the probe to full depth. Alignment springs position the probe within the casing and assure a well-defined radiative coupling between the probe and the borehole. Radiation shields on the hole prevent direct sunlight from reaching the bottom of the hole.

As a part of the field geology experiment, the astronaut will drill a third hole near the HFE, using corestem drill sections and a coring bit, and obtain cores of lunar material for subsequent analysis of thermal properties and composition. Total available core length is 2.5 m (100 in.).

The heat flow experiment, design, and data analysis are the responsibility of Dr. Marcus Langseth of the Lamont-Doherty Geological Observatory.

Active Seismic Experiment: The ASE will produce data on the physical structure and bearing strength of the lunar surface by measuring seismic waves. Two types of man-made seismic sources will be used with the ASE: a crew-actuated pyrotechnic "thumper" and a mortar-like device from which four rocket propelled projectiles can be launched by command from Earth. Naturally produced seismic events will be detected passively by the ASE (the ASE geophones will be turned on remotely for short listening periods). The seismic waves are detected by geophones deployed by the crew. Data on wave penetration, frequency spectra, and velocity to lunar depths of 152 meters (500 ft.) will be obtained and passed to the ALSEP central station for transmittal to the Earth. Dr. Robert Kovach of Stanford University is the Principal Investigator.

The mortar like device will be deployed, aligned and activated about 15 meters (48 ft.) north of the ALSEP central station. The four grenade-like projectiles will be launched sometime after the crew returns and at a time specified by the principal investigator, but no sooner than the third lunation.

The crew will deploy three geophones at 3.5, 45.7 and 91.4 meters (10, 150, 300 ft.) from the ALSEP central station. Enroute back to the central station, the crewman will fire 19 "thumper" charges at 4.5 meters (15 ft.) intervals along the geophone line. The thumper serves as a storage and transport rack for the geophones and their connecting cable.

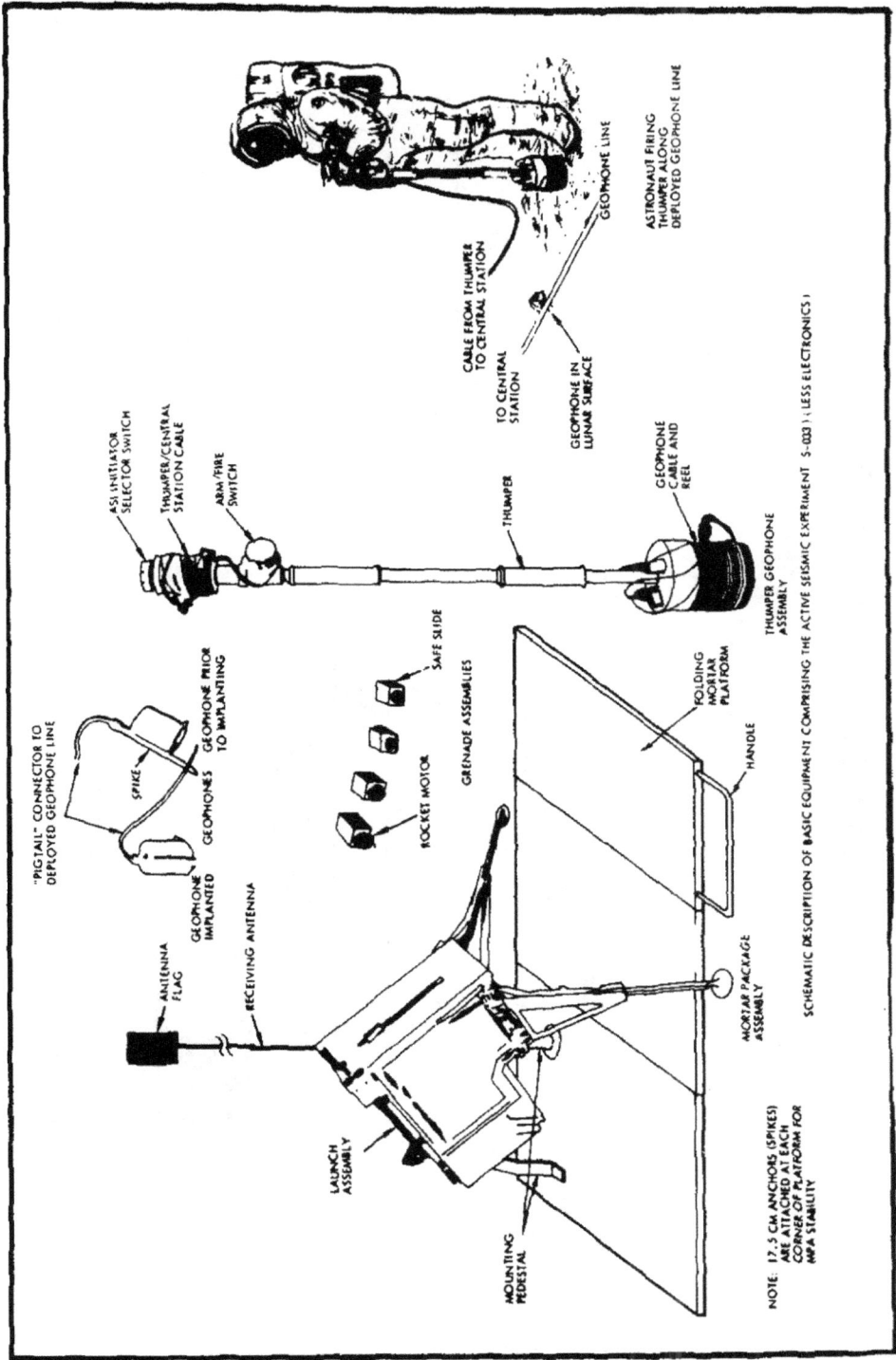

Active Seismic Experiment

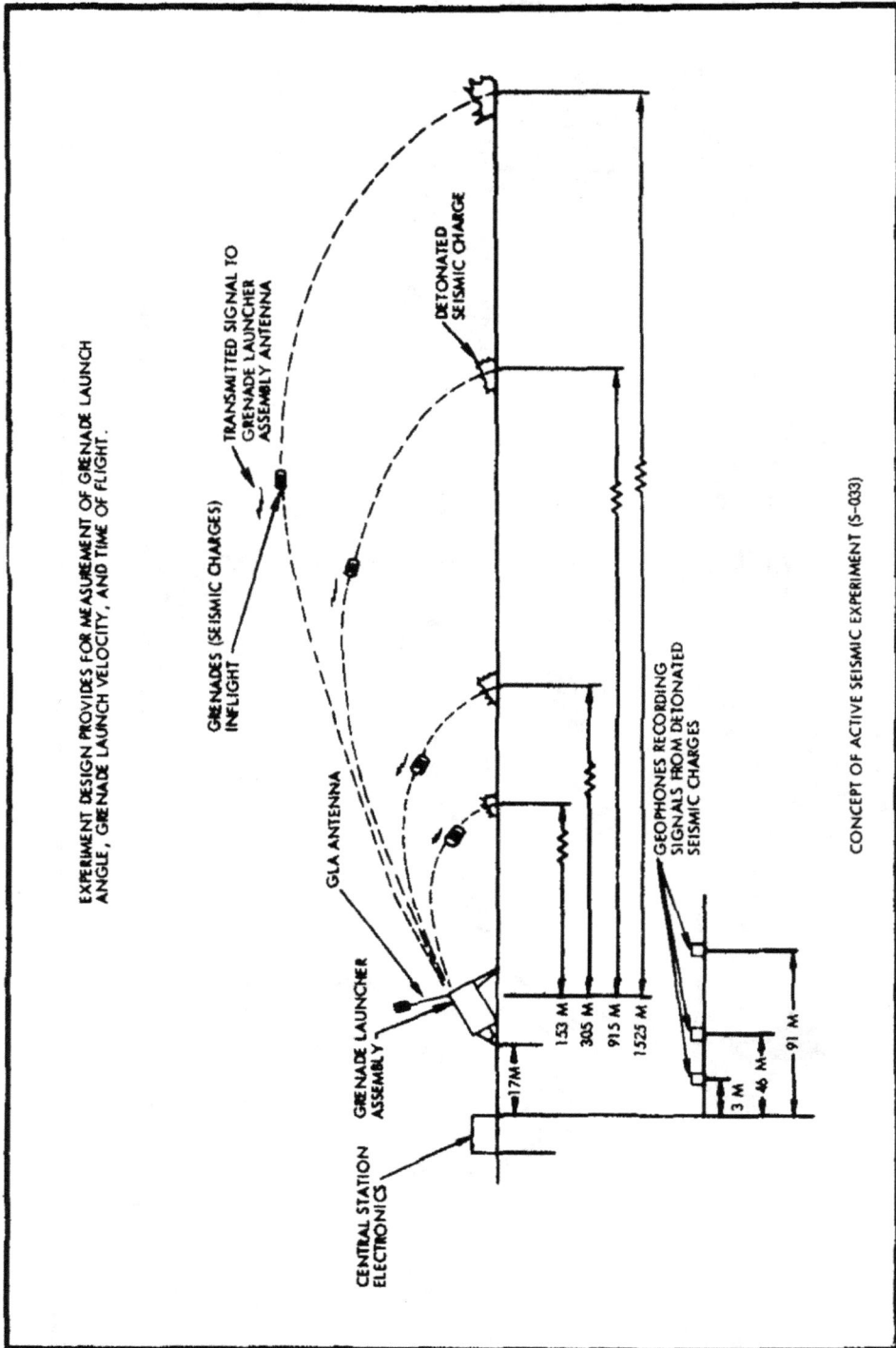

Active Seismic Experiment Mortar Mode Concept

The two major components of the ASE are:

1. The thumper-geophone assembly measuring 1.13 m
(44.5 in.) when deployed and weighing 3.2 kg. (7 lbs.) including three geophones and cable. Each geophone is 12.2 cm
(4.8 in.) high, 4.1 cm (1.6 in.) in diameter and weighs less
than 0.4 kg. (1 lb.).

2. The package projectile launch assembly weighs 6.8 kg.
(15 lbs.) (including four projectiles) and is 24 cm (9.5 in.)
high, 10.2 cm (4 in.) wide and 38 cm (15.6 in.) long. The
mortar-like launching device is made of fiberglass and magnesium, and contains firing circuitry and a receiver antenna.
The projectile launch assembly is enclosed in a box and consists of four fiberglass launch tubes and four projectiles.
The projectiles vary in length and weight according to the
propellant and explosive charges. Radio transmitters in
each projectile furnish start-and-stop flight time data for
telemetry back to Earth. Thus, with the launch angle known,
range can be calculated. The geophones provide information
on seismic wave travel time. Correlation of this time with
range will establish wave velocity through the lunar surface.

ALSEP Central Station: The central station serves as a pow-
er-distribution and data-handling point for experiments carried on the ALSEP. Central station components are the data
subsystem, helical antenna, experiment electronics, and power conditioning unit. The central station is deployed after
other experiment instruments are unstowed from the pallet.

The central station data subsystem receives and decodes
uplink commands, times and controls experiments, collects
and transmits scientific and engineering data downlink, and
controls the electrical power subsystem through the power
distribution and signal conditioner.

The modified axial-helix S-band antenna receives and
ansmits a right-hand circularly-polarized signal. The
antenna is manually aimed with a two-gimbal azimuth/elevation
aiming mechanism.

The ALSEP electrical power subsystem draws electrical
power from a SNAP-27 (Systems for Nuclear Auxiliary Power)
radioisotope thermoelectric generator.

-more-

SNAP-27--Power Source for ALSEP

A SNAP-27 unit, similar to three others deployed on the Moon, will provide power for the ALSEP package. SNAP-27 is one of a series of radioisotope thermoelectric generators, or atomic batteries, developed by the Atomic Energy Commission under its space SNAP program. The SNAP (Systems for Nuclear Auxiliary Power) program is directed at development of generators and reactors for use in space, on land, and in the sea.

SNAP-27 on Apollo 12 marked the first use of a nuclear power system on the Moon. The nuclear generators are required to provide power for periods of at least one year. Thus far, the SNAP-27 unit on Apollo 12 has operated over two years; and the SNAP-27 unit on Apollo 14 has operated a little over a year.

The basic SNAP-27 unit is designed to produce at least 68 watts of electrical power. The SNAP-27 unit is a cylindrical generator, fueled with the radioisotope plutonium-238. It is about 46 cm (18 inches) high and 41 cm (16 inches) in diameter, including the heat radiating fins. The generator, making maximum use of the lightweight material beryllium, weighs about 12.7 kilograms (28 pounds) without fuel.

The fuel capsule, made of a superalloy material, is 42 cm (16.5 inches) long and 6.4 cm (2.5 inches) in diameter. It weighs about 7 km (15.5 pounds), of which 3.8 km (8.36 pounds) represent fuel. The plutonium-238 fuel is fully oxidized and is chemically and biologically inert.

The rugged fuel capsule is stowed within a graphite fuel cask from launch through lunar landing. The cask is designed to provide reentry heating protection and containment for the fuel capsule in the event of an aborted mission. The cylindrical cask with hemispherical ends includes a primary graphite heat shield, a secondary beryllium thermal shield, and a fuel capsule support structure. The cask is 58.4 cm (23 inches) long and 20 cm (eight inches) in diameter and weighs about 11 km (24.5 pounds). With the fuel capsule installed, it weighs about 18 km (40 pounds). It is mounted on the lunar module descent stage.

-more-

Once the lunar module is on the Moon, an Apollo astronaut will remove the fuel capsule from the cask and insert it into the SNAP-27 generator which will have been placed on the lunar surface near the module.

The spontaneous radioactive decay of the plutonium-238 within the fuel capsule generates heat which is converted directly into electrical energy--at least 68 watts. Units now on the lunar surface are producing 72 to 74 watts. There are no moving parts.

The unique properties of plutonium-238, make it an excellent isotope for use in space nuclear generators. At the end of almost 90 years, plutonium-238 is still supplying half of its original heat. In the decay process, plutonium-238 emits mainly the nuclei of helium (alpha radiation), a very mild type of radiation with a short emission range.

Before the use of the SNAP-27 system was authorized for the Apollo program a thorough review was conducted to assure the health and safety of personnel involved in the launch and of the general public. Extensive safety analyses and tests were conducted which demonstrated that the fuel would be safely contained under almost all credible accident conditions.

Soil Mechanics: Mechanical properties of the lunar soil, surface, and subsurface will be investigated through trenching at various locations and through use of the self-recording penetrometer. The self-recording penetrometer measures the characteristics and mechanical properties of the lunar surfac material. The penetrometer consists of a 76 cm (30-in) penetration shaft and recording drum. Three interchangeable penetration cones 1.3, 3.2 and 6.4 cm^2 (0.2, 0.5 and 1.0 in)^2 cross sections and a 2.5 x 12.7 cm (1 x 5-in.) pressure plate may be attached to the shaft. The crewman forces the penetrometer into the surface and a stylus scribes a force vs. depth plot on the recording drum. The drum can record up to 24 force-depth plots. The upper housing containing the recording drum is detached at the conclusion of the experiment for return to Earth and analysis by the principal investigator. This experiment will be documented with the electric Hasselblad and the 16mm data acquisition cameras.

Soil behavior characteristics are also determined from interactions of LRV wheels, LM footpads, and footprints with the lunar soil.

Lunar Portable Magnetometer (LPM): The Apollo 16 crew will use the LPM during EVAs 1 and 3 for measuring variations in the lunar magnetic field at several points during the geology traverses. Data gathered will be used to determine the location, strength, and dimensions of the magnetic field, as well as knowledge of the local selenological structure. The LPM experiment was also flown on Apollo 14.

The LPM will be carried on the LRV aft pallet and consists of a flux-gate magnetometer sensor head mounted on a tripod and an electronics data package. The sensor head is connected to the data package by a 15.2-meter (50 ft.) flat cable, and after the crewman aligns the sensor head at least 14 meters (46 ft.) from the data package, he returns to the LRV and relays readouts to Earth by voice.

The sensor head is sensitive to magnetic interference from the crew's portable life support systems, the geology tool, and the LRV, hence the need for emplacing and aligning the sensor well away from them. The mercury-cell powered electronics package has a range of ±256 gamma. Readings are displayed in three digital meters---one for each axis (orthogonal X, Y and Z). At each traverse location for LPM measurements, the crew will call out the meter readings on each axis.

-more-

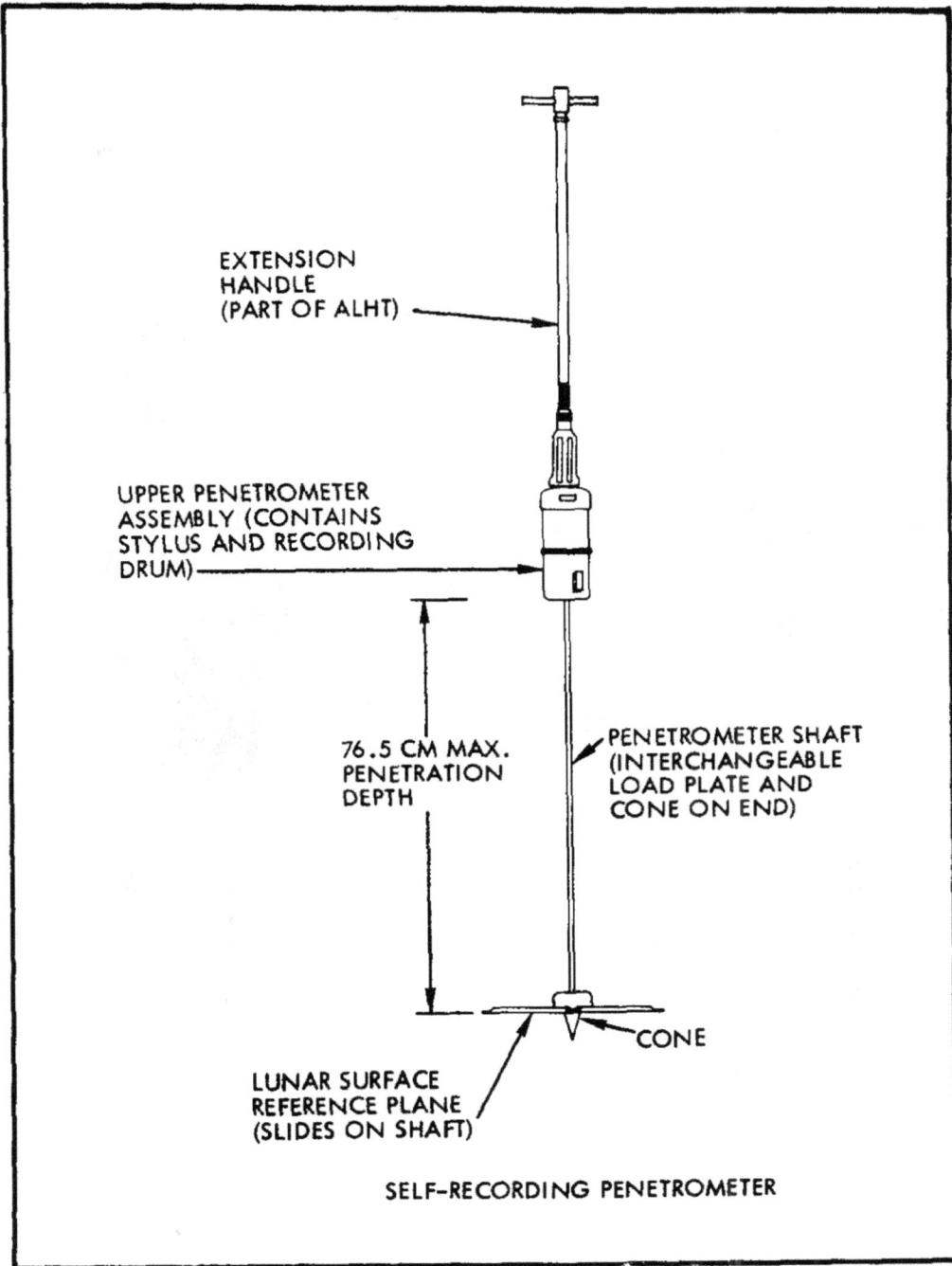

EXTENSION
HANDLE
(PART OF ALHT)

UPPER PENETROMETER
ASSEMBLY (CONTAINS
STYLUS AND RECORDING
DRUM)

76.5 CM MAX.
PENETRATION
DEPTH

PENETROMETER SHAFT
(INTERCHANGEABLE
LOAD PLATE AND
CONE ON END)

CONE

LUNAR SURFACE
REFERENCE PLANE
(SLIDES ON SHAFT)

SELF-RECORDING PENETROMETER

Soil Mechanics Experiment Penetrometer

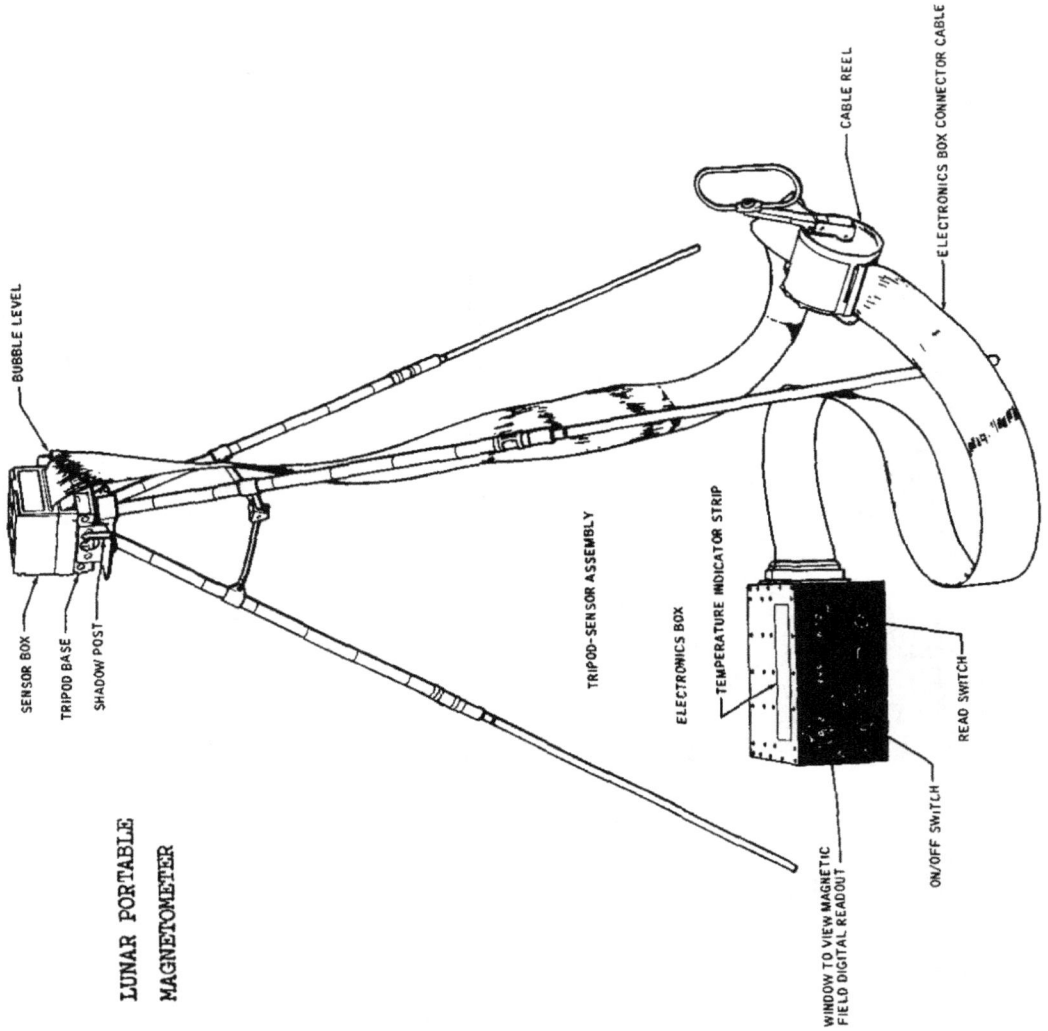

LUNAR PORTABLE
MAGNETOMETER

BUBBLE LEVEL

SENSOR BOX

TRIPOD BASE

SHADOW POST

TRIPOD-SENSOR ASSEMBLY

CABLE REEL

ELECTRONICS BOX CONNECTOR CABLE

ELECTRONICS BOX

TEMPERATURE INDICATOR STRIP

READ SWITCH

ON/OFF SWITCH

WINDOW TO VIEW MAGNETIC
FIELD DIGITAL READOUT

Dimensions of the LPM components are 10.2 x 19 x 12.7 cm (4 x 7 1/2 x 5 inches) for the data package and 8.6 x 14.4 x 6.7 cm (3 3/8 x 5 11/16. x 2 5/8 inches) for the sensor head. The sensor head tripod is 45.7 cm (18 inches) long when retracted and extends to 78.7 cm (31 inches). The principal investigator is Dr. Palmer Dyal, NASA Ames Research Center, Mountain View, Calif.

Far Ultraviolet Camera/Spectroscope: Deep-space concentrations of hydrogen in interplanetary, interstellar and intergalactic regions will be mapped by this experiment using an instrument which gathers both photographic images and spectroscopic data in the far ultraviolet spectrum. The experiment will be the first such astronomical observation emplaced on the lunar surface.

Earlier spectrographic searches for hydrogen sources in space from Earth-orbiting astronomical satellites were impaired by the "masking" effect of the Earth's corona. The instrument will be pointed toward such targets as star clouds, nebulae, galaxy clusters and other galactic objects, intergalactic hydrogen, the solar bow cloud and solar wind, lunar atmosphere and any possible lunar volcanic gases, and the Earth's atmosphere and corona. Several astrophysicists have speculated that extremely hot hydrogen exists in intergalactic space and that hydrogen clusters may be detected between galaxies by an instrument such as the far UV camera spectroscope.

The experiment includes a 75mm (3-in) electronographic Schmidt camera with a potassium bromide cathode and 35mm film magazine and transport. Emplaced on a tripod in the LM shadow, the instrument is provided with a battery pack which is placed in the Sun at the end of its connecting cable. The spectroscope is fitted with lithium fluoride and calcium fluoride filters for detecting hydrogen Lyman-alpha radiation in the 1216 Å (angstrom) wavelength.

Measurements will range from 500 to 1550 Å for spectroscopic data and from 1050 to 1550 Å and 1230 to 1550 Å in photographic imagery. Hydrogen gas clouds will be detected through differential measurements of the photoimagery.

Using the elevation/azimuth adjustments on the instrument mount, the crew will align the camera/spectroscope toward specific targets periodically during the three EVAs. Near the end of EVA 3, the camera film cassette will be retrieved for return to Earth.

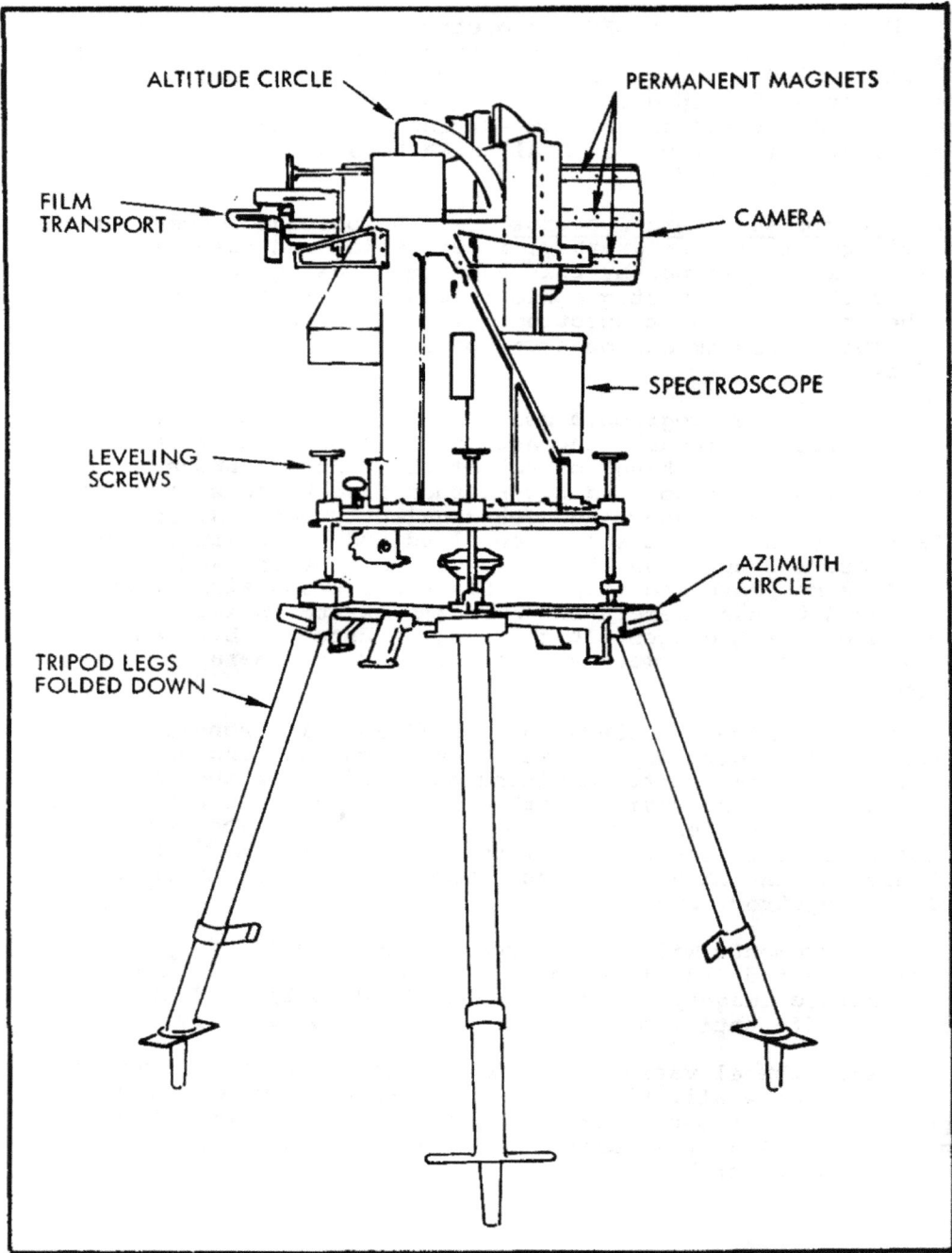

Tripod-Mounted Schmidt Electronographic UV Camera/Spectroscope

Experiment principal investigator is Dr. George R. Carruthers of the E.O. Hurlburt Center for Space Research, Naval Research Laboratory, Washington, D.C.

Solar Wind Composition Experiment (SWC): The scientific objective of the solar wind composition experiment is to determine the elemental and isotopic composition of the noble gases in the solar wind.

As in Apollos 11, 12, 14 and 15, the SWC detector will be deployed on the lunar surface and brought back to Earth by the crew. The detector will be exposed to the solar wind flux for 46 hours compared to 41 hours on Apollo 15, 21 hours on Apollo 14, 18 hours on Apollo 12 and two hours on Apollo 11.

The solar wind detector consists of an aluminum and platinum foil .37 square meters (4 square ft.) in area and about 0.5 mils thick rimmed by Teflon for resistance to tearing during deployment. The platinum foil strips have been added to the experiment on Apollo 16 to reduce contamination. This will allow a more accurate determination of solar wind constituents. A staff and yard arrangement will be used to deploy the foil and to maintain the foil approximately perpendicular to the solar wind flux. Solar wind particles will penetrate into the foil, while cosmic rays pass through. The particles will be firmly trapped at a depth of several hundred atomic layers. After exposure on the luanr surface, the foil is rolled up and returned to Earth. Professor Johannes Geiss, University of Berne, Switzerland, is principal investigator. Professor Geiss is sponsored and funded by the Swiss Committee for Space Research.

Cosmic Ray Detector

A four-panel detector array is mounted on the lunar module descent stage for measuring the charge, mass and energy of cosmic ray and solar wind particles impacting the detector array on the way to the Moon and during the stay on the lunar surface. The detector array will be brought back to Earth for analysis.

Energy ranges for the two types of particles to be measured are from 0.5 to 10 thousand electron volts/nucleon and from 0.2 to 200 million electron volts/nucleon. Additionally, various types of glass detectors will be evaluated and lunar surface thermal neutron flux will be measured.

Cosmic ray and solar wind particles will impact three portions of the detector during translunar coast, and the fourth panel will be uncovered early in the first lunar surface EVA to gather data on lunar cosmic ray and solar wind particles. The detector will be folded and bagged during the third EVA and returned to Earth.

Detector panel 1 is a sandwich of 31 sheets of Lexan, each 0.025 cm thick, and covered by perforated aluminized Teflon.

Panel 2 is almost identical to panel 1, except that two plastic sheets are pre-irradiated for data analysis calibration.

Panel 3 is made up of 40 layers of 0.2-cm-thick Kodacel cellulose triacetate sheets, overlaid on the upper half by 10 five-micron thick Lexan sheets. Five specimens of glass are imbedded in the panel's lower portion.

Panel 4 is similar in makeup to panel 3, except that the sheets in the lower part of the panel may be shifted. A 0.076-cm-thick aluminum plate, to which are bonded small fragments of mica, glass and natural crystals, covers the lower quarter of the panel. The next highest quarter of the panel is covered by a bonded 2-micron thick foil. A lanyard-actuated slide made of 0.0005-cm thick aluminum foil bonded to a platinum base covers the upper half of panel four. The panel upper half is exposed to the lunar environment when the crewman pulls the lanyard during EVA 1.

Principal investigators for the experiment are Dr. R. L. Fleischer, General Electric Co., Dr. P. B. Price, University of California at Berkely, and Dr. R. M. Walker, Washington University.

Lunar Geology Investigation

The fundamental objective of the lunar geology investigation experiment is to provide data in the vicinity of the landing site for use in the interpretation of the geologic history of the Moon. Apollo lunar landing missions offer the opportunity to correlate carefully collected samples with a variety of observational data on at least the upper portions of the mare basin filling and the lunar highlands, the two major geologic subdivisions of the Moon. The nature and origin of the maria and highlands will bear directly on the history of lunar differentiation and differentiation processes. From the lunar bedrock, structure, land forms and special materials, information will be gained about the internal processes of the Moon. The nature and origin of the debris layer (regolith) and the land forms superimposed on the maria and highland regions are a record of lunar history subsequent to their formation. This later history predominately reflects the history of the extra-lunar environment. Within and on the regolith, there will also be materials that will aid in the understanding of geologic units elsewhere on the Moon and the broader aspects of lunar history.

The primary data for the lunar geology investigation experiment come from photographs, verbal data, and returned lunar samples. Photographs taken according to specific procedures will supplement and illustrate crew comments, record details not discussed by the crew, provide a framework for debriefing, and record a wealth of lunar surface information that cannot be returned or adequately described by any other means.

In any Hasselblad picture taken from the lunar surface, as much as 90 percent of the total image information may be less than 100 feet from the camera, depending on topography and how far the camera is depressed below horizontal. Images of distant surface detail are so foreshortened that they are difficult to interpret. Therefore, it is important that panoramas be taken at intervals during the traverse and at the farthest excursion of the traverse. This procedure will extend the high resolution photographic coverage to the areas examined and discussed by the astronaut, and will show the regional context of areas of specific interest that have been discussed and photographed in detail.

- more -

The polarizing filters will permit the measurement of the degree of polarization and orientation of the plane of polarization contained in light reflected from the lunar surface. Different lunar materials (i.e., fine-grained glass and/or fragments, strongly shocked rocks, slightly shocked rocks and shock-lithified fragmental material) have different polarimetric functions, in other words, different polarimetric "signatures." Comparison of the polarimetric function of known material, such as returned samples and close-up lunar surface measurements, with materials photographed beyond the traverse of the astronaut will allow the classification and correlation of these materials even though their textures are not resolvable. The polarimetric properties of lunar materials and rock types are a useful tool for correlation and geologic mapping of each landing site, and for extrapolation of geologic data from site to site across the lunar surface.

The "in situ" photometric properties of both fine-grained materials and coarse rock fragments will serve as a basis for delineating, recognizing, describing, and classifying lunar materials. The gnomon, with photometric chart attached, will be photographed beside a representative rock and, if practical, beside any rock or fine-grained material with unusual features.

The long focal length (500 mm) lens with the HEDC will be used to provide high resolution data. A 5 to 10 centimeter resolution is anticipated at a distance of 1 to 2 km (0.6 to 1.2 mi.). The high degree of resolution will make it possible to analyze the stratigraphic layering in North Ray Crater walls and potential outcrops and exposed bedrock in distant areas.

Small exploratory trenches, several centimeters deep, are to be dug to determine the character of the regolith down to these depths. The trenches should be dug in the various types of terrain and in areas where the surface characteristics of the regolith are of significant interest as determined by the astronaut crew. The main purpose of the trenches will be to determine the small scale stratigraphy (or lack of) in the upper few inches of the regolith in terms of petrological characteristics and particle size.

- more -

The organic control sample, carried in each sample return case (SRC) will be analyzed after the mission in the Lunar Receiving Laboratory to determine the level of contamination in each SRC. This will then be compared to an organic control sample which was removed from the SRC prior to the SRC being shipped to KSC for loading onto the LM.

In order to more fully sample the major geological features of the Apollo 16 landing site, various groupings of sampling tasks are combined and will be accomplished in concentrated areas. This will aid in obtaining vertical as well as lateral data to be obtained in the principal geological settings. Thus, some trench samples, core tube samples and lunar environmental soil samples will be collected in association with comprehensive samples. In addition, sampling of crater rims of widely differing sizes in a concentrated area will give a sampling of the deeper stratigraphic divisions at that site. Repeating this sampling technique at successive traverse stations will show the continuity of the main units within the Cayley formation.

Sampling and photographic techniques used to gather data in the Descartes region include:

* Documented samples of lunar surface material which, prior to gathering, are photographed in color and stereo -- using the gnomon and photometric chart for comparison of position and color properties -- to show the sample's relation to other surface features.

* Rock and soil samples of Imbrian age rocks from deep layers, and soil samples from the regolith in the immediate area where the rocks are gathered.

* Radial sampling of material on the rim of a fresh crater -- material that should be from the deepest strata.

* Photopanoramas for building mosaics which will allow accurate control for landing site map correlation.

* Polarimetric photography for comparison with known materials.

* Double drive tube samples to depths of 60 cm (23.6 in.) for determining the stratigraphy in multi-layer areas.

- more -

* Single drive tube samples to depths of 38 cm (15 in.) in the comprehensive sample area and in such target of opportunity areas as mounds and fillets.

* Drill core sample of the regolith which will further spell out the stratigraphy of the area sampled.

* A small exploratory trench, ranging from 6 to 24 cm (2.4 to 9.7 in.) in depth, to determine regolith particle size and small-scale stratigraphy.

* Large equidimensional rocks ranging from 15 to 24 cm (6 to 9.4 in.) in diameter for data on the history of solar radiation. Similar sampling of rocks from 6 to 15 cm (2.4 to 6 in.) in diameter will also be made.

* Fillet sampling which may reveal a relationship between fillet volume and rock size to the length of time the fillet has been in place.

* Vacuum-packed lunar environment soil and rock samples kept biologically pure for postflight gas, chemical and microphysical analysis.

* Special surface samples to determine properties of the upper 10 - 100 microns of the lunar surface.

- more -

Apollo Lunar Geology Hand Tools

Sample scale - The scale is used to weigh the loaded sample
return containers, sample bags, and other containers to
maintain the weight budget for return to Earth. The scale
has graduated markings in increments of 5 pounds to a
maximum capacity of 80 pounds. The scale is stowed and
used in the lunar module ascent stage.

Tongs - The tongs are used by the astronaut while in a
standing position to pick up lunar samples from pebble
size to fist size. The tines of the tongs are made of
stainless steel and the handle of aluminum. The tongs
are operated by squeezing the T-bar grips at the top of
the handle to open the tines. In addition to picking up
samples, the tongs are used to retrieve equipment the
astronaut may inadvertantly drop. This tool is 81 cm (32
inches) long overall.

Lunar rake - The rake is used to collect discrete samples
of rocks and rock chips ranging from 1.3 cm (one-half inch)
to 2.5 cm (one inch) in size. The rake is adjustable for
ease of sample collection and stowage. The tines, formed
in the shape of a scoop, are stainless steel. A handle,
approximately 25 cm (10 inches) long, attaches to the extension handle for sample collection tasks.

Adjustable scoop - The sampling scoop is used to collect
soil material or other lunar samples too small for the rake
or tongs to pick up. The stainless steel pan of the scoop,
which is 5 cm (2 inches) by 11 cm (41/2 inches) by 15 cm
(6 inches) has a flat bottom flanged on both sides and a
partial cover on the top to prevent loss of contents. The
pan is adjustable from horizontal to 55 degrees and 90
degrees from the horizontal for use in scooping and
trenching. The scoop handle is compatible with the extension
handle.

Hammer - This tool serves three functions; as a sampling
hammer to chip or break large rocks, as a pick, and as a
hammer to drive the drive tubes or other pieces of lunar
equipment. The head is made of impact resistant tool steel,
has a small hammer face on one end, a broad flat blade on the
other, and large hammering flats on the sides. The handle,
made of aluminum, is 36 cm (14 inches) long; its lower
end fits the extension handle when the tool is used as a hoe.

-more-

Extension handle - The extension handle extends the astronaut's reach to permit working access to the lunar surface by adding 76 cm (30 inches) of length to the handles of the scoop, rake, hammer, drive tubes, and other pieces of lunar equipment. This tool is made of aluminum alloy tubing with a malleable stainless steel cap designed to be used as an anvil surface. The lower end has a quick-disconnect mount and lock designed to resist compression, tension, torsion, or a combination of these loads. The upper end is fitted with a sliding "T" handle to facilitate any torqueing operation.

Drive Tubes - These nine tubes are designed to be driven or augured into soil, loose gravel, or soft rock such as pumice. Each is a hollow thin-walled aluminum tube 41 cm (16 inches) long and 4 cm (1.75 inch) diameter with an integral coring bit. Each tube can be attached to the extension handle to facilitate sampling. A deeper core sample can be obtained by joining tubes in series of two or three. When filled with sample, a Teflon cap is used to seal the open end of the tube, and a keeper device within the drive tube is positioned against the top of the core sample to preserve the stratigraphic integrity of the core. Three Teflon caps are packed in a cap dispenser that is approximately a 5.7 cm (2.25 inch) cube.

Gnomon and Color Patch - The gnomon is used as a photographic reference to establish local vertical Sun angle, scale, and lunar color. This tool consists of a weighted staff mounted on a tripod. It is constructed in such a way that the staff will right itself in a vertical position when the legs of the tripod are on the lunar surface. The part of the staff that extends above the tripod gimbal is painted with a gray scale from 5 to 35 percent reflectivity and a color scale of blue, orange, and green. The color patch, similarly painted in gray scale and color scale, mounted on one of the tripod legs provides a larger target for accurately determining colors in color photography.

-more-

Apollo Lunar Hand Tool Carrier - This is an aluminum framework with a handle and legs that unfold to provide a steady base. The tools described above may be stowed in special fittings on this framework during lunar surface EVAs. The framework also provides support positions for two sample collection bags and a 20-bag documented sample bag dispenser. The hand tool carrier mounts on the Quad III LRV pallet during launch and stays on that pallet when it is transferred to the aft end of the lunar roving vehicle.

Sample Bags - Several different types of bags are furnished for collecting lunar surface samples. The Teflon documented sample bag (DSB), 19 by 20 cm (7-1/2 by 8 inches) in size, is prenumbered and packed in a 20-bag dispenser that can be mounted on the hand tool carrier or on a bracket on the Hasselblad camera. Documented sample bags (140) will be available during the lunar surface EVAs. The sample collection bag (SCB), also of Teflon, has interior pockets along one side for holding drive tubes and exterior pockets for the special environmental sample container and for a drive tube cap dispenser. This bag is 17 by 23 by 41 cm (6-3/4 by 9 by 16 inches) in size (exclusive of the exterior pockets) and fits inside the sample return containers. During the lunar surface EVAs this bag is hung on the hand tool carrier or on the portable life support system tool carrier. Four SCBs will be carried on Apollo 16. The extra sample collection bag (ESCB) is identical to the SCB except that the interior and exterior pockets are omitted. During EVAs it is handled in the same way as an SCB. Four ESCB bags will be carried on the mission. A sample return bag, 13 by 33 by 57 cm (5 by 13 by 22.5 inches) in size, replaces the third sample return container and is used for the samples collected on the third EVA. It hangs on the LRV pallet during this EVA.

A special type of sample bag, the padded sample bag, is carried on Apollo 16. Although generally similar to the DSB, it has an inner padding of knit Teflon that forms an open-topped box 6 cm (2-1/2 inches) thick and a velcro strap to insure satisfactory closure for return to Earth. Only two of these special purpose bags will be carried on the flight.

Sample Return Container - This container maintains a vacuum environment and padded protection for lunar samples. It is an aluminum box 20 by 28 by 48 cm (8 by 11.5 by 19 inches) in size, with a knife-edge indium-silver seal, a strap latch system for closing, and a lever and pin system to support the container in the LM and CM stowage compartments under all vibration and g-force conditions. The drive tubes, special environmental sample container, and the core sample vacuum container as well as some other specialized pieces of equipment are flown outbound and inbound within the sample return containers, two of which will be carried on Apollo 16.

Special Environmental Sample Containers - This container is used to protect the vacuum environment of selected samples of lunar soil or rocks to be studied in specific experiments upon return to Earth. It is a thin-walled stainless steel can with a knife edge at the top and three-legged press assembly attached to the lid to effect a vacuum-proof seal when used on the lunar surface. Until used upon collection of the sample, the seal surfaces are covered with Teflon protectors that are then discarded.

Core Sample Vacuum Container - The core sample vacuum container is a receptacle for vacuum storage and transport to Earth of a lunar surface drive tube. The core sample vacuum container is about twice as long as the special environmental sample container and includes an insert to grip the drive tube, providing lateral and longitudinal restraint.

Surface Sampler Tool - The purpose of the surface sampler tool is to capture and return to Earth for analysis a sample from the very top of the lunar surface. The surface sampler tool hopefully will capture 300 to 500 milligrams of material only 100 to 500 microns deep. Each sampler consists of a 1.25-cm (1/2-inch) deep box which attaches to the universal hand tool. The 11.25-by-12.5-cm (2.5-by-5-inch) boxes each contain a plate which is floating in channels inside the box. One plate is covered with a deep pile velvet cloth and the other plate will be covered with either beta cloth or a thin layer of grease.

The astronaut will open the spring-loaded door at the box bottom and gently lower the box until the floating plate touches the lunar surface. The lunar surface grains will be trapped in the fabric pile or grease. The astronaut will lift the box, close the lid, and place the box in a numbered bag for return to Earth.

Lunar Geology Hand Tools

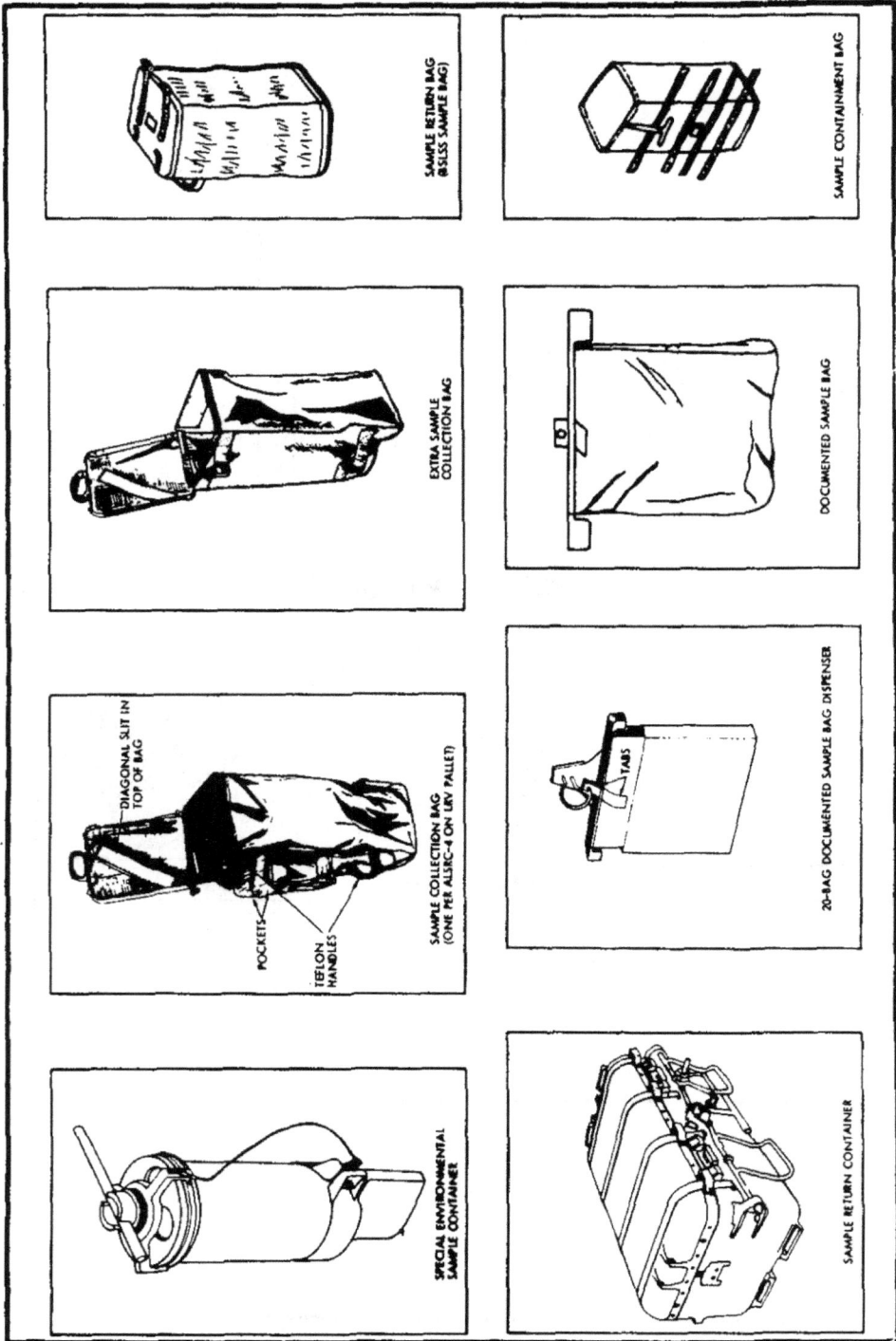

SAMPLE RETURN BAG (BSLSS SAMPLE BAG)

SAMPLE CONTAINMENT BAG

EXTRA SAMPLE COLLECTION BAG

DOCUMENTED SAMPLE BAG

DIAGONAL SLIT IN TOP OF BAG

SAMPLE COLLECTION BAG (ONE PER ALSRC-4 ON LRV PALLET)

POCKETS

TEFLON HANDLES

TABS

20-BAG DOCUMENTED SAMPLE BAG DISPENSER

SPECIAL ENVIRONMENTAL SAMPLE CONTAINER

SAMPLE RETURN CONTAINER

Lunar Geology Sample Containers

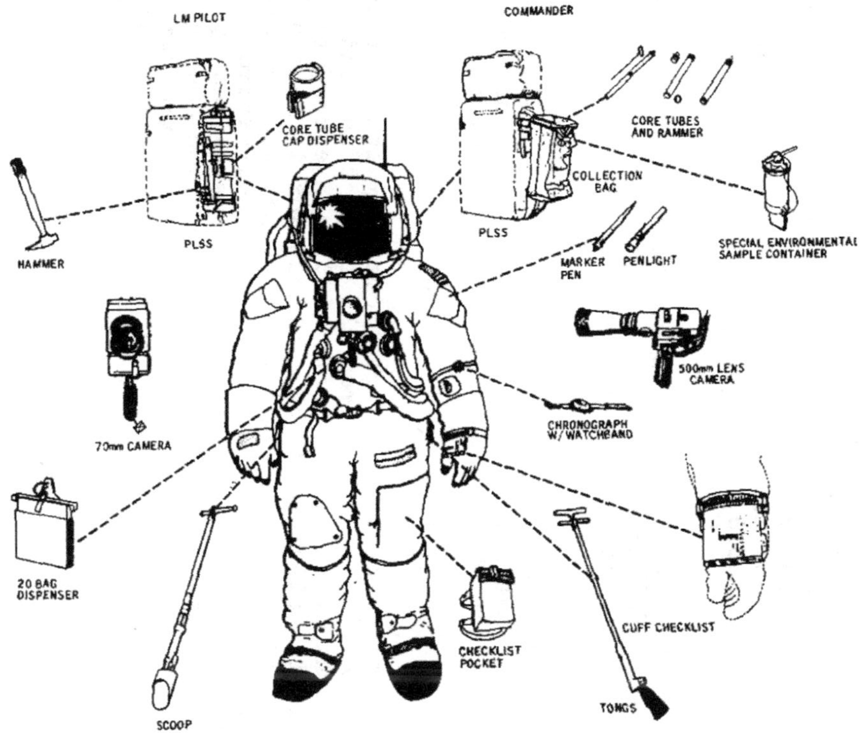

EXTRACTOR

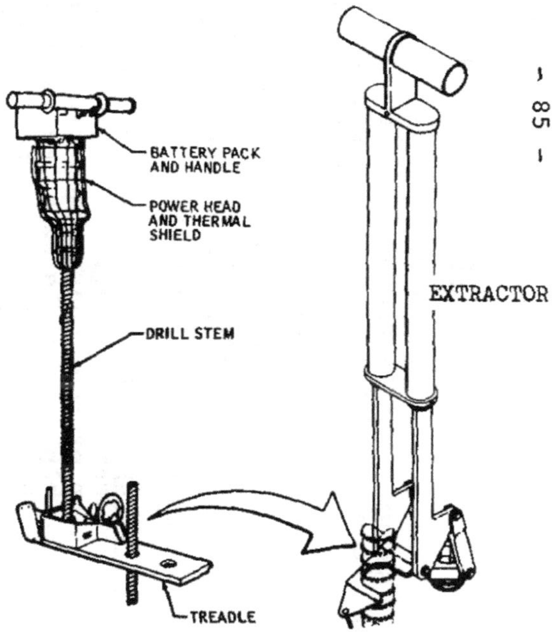

BATTERY PACK AND HANDLE

POWER HEAD AND THERMAL SHIELD

DRILL STEM

TREADLE

COMMANDER

CORE TUBES AND RAMMER

SPECIAL ENVIRONMENTAL SAMPLE CONTAINER

COLLECTION BAG

PENLIGHT

MARKER PEN

500mm LENS CAMERA

CHRONOGRAPH W/ WATCHBAND

CUFF CHECKLIST

TONGS

PLSS

CHECKLIST POCKET

CORE TUBE CAP DISPENSER

LM PILOT

PLSS

70mm CAMERA

HAMMER

20 BAG DISPENSER

SCOOP

LUNAR ORBITAL SCIENCE

Service Module Sector 1 houses the Scientific Instrument Module (SIM) bay. Eight experiments are carried in the SIM bay: X-ray fluorescence spectrometer, gamma ray spectrometer, alpha-particle spectrometer, panoramic camera, 76-millimeter (3-inch) mapping camera, laser altimeter and a mass spectrometer; a subsatellite carries three integral experiments (particle detectors, magnetometer and S-band transponder) comprising the eighth SIM bay experiment and will be placed into lunar orbit.

Gamma-Ray Spectrometer: On a 7.6 meter (25-foot) extendable boom, the gamma-ray spectrometer measures the chemical composition of the lunar surface in conjunction with the X-ray and alpha-particle experiments to gain a compositional "map" of the lunar surface ground track. It detects natural and cosmic rays, induced gamma radioactivity and will operate on the Moon's dark and light sides. Additionally, the experiment will be extended in transearth coast to measure the radiation flux in cislunar space and record a spectrum of cosmological gamma-ray flux. The device can measure energy ranges between 0.1 to 10 million electron volts. The extendable boom is controllable from the command module cabin. Principal investigator is Dr. James R. Arnold, University of California at San Diego.

X-Ray Fluorescence Spectrometer: This geochemical experiment measures the composition of the lunar surface from orbit, and detects X-ray fluorescence caused by solar X-ray interaction with the Moon. It will analyze the sunlit portion of the Moon. The experiment will measure the galactic X-ray flux during transearth coast. The device shares a compartment on the SIM bay lower shelf with the alpha-particle experiment, and the protective door may be opened and closed from the command module cabin. Principal investigator is Dr. Isidore Adler, NASA Goddard Space Flight Center, Greenbelt, Md.

Alpha-Particle Spectrometer: This spectrometer measures mono-energetic alpha particles emitted from the lunar crust and fissures as products of radon gas isotopes in the energy range of 4.7 to 9.3 million electron volts. The sensor is made up of an array of 10 silicon surface barrier detectors. The experiment will consturct a "map" of lunar surface alpha-particle emissions along the orbital track and is not constrained by solar illumination. It will also measure deep-space alpha-particle background emissions in lunar orbit and in transearth coast. Protective door operation is controlled from the cabin. Principal investigator is Dr. Paul Gorenstein, American Science and Engineering, Inc., Cambridge, Mass.

	APOLLO MISSIONS					
	11	12	14	15	16	17
SERVICE MODULE EXPERIMENTS						
S-160 GAMMA-RAY SPECTROMETER				X	X	
S-161 X-RAY SPECTROMETER				X	X	
S-162 ALPHA-PARTICLE SPECTROMETER				X	X	
S-164 S-BAND TRANSPONDER (CSM/LM)			X	X	X	X
S-165 MASS SPECTROMETER				X	X	
S-169 FAR UV SPECTROMETER						X
S-170 BISTATIC RADAR			X	X	X	
S-171 IR SCANNING RADIOMETER						X
S-209 LUNAR SOUNDER						X
SUBSATELLITE:						
S-164 S-BAND TRANSPONDER				X	X	X
S-173 PARTICLE MEASUREMENT				X	X	X
S-174 MAGNETOMETER				X	X	X
SM. PHOTOGRAPHIC TASKS:						
24" PANORAMIC CAMERA				X	X	X
3" MAPPING CAMERA				X	X	X
LASER ALTIMETER				X	X	X
COMMAND MODULE EXPERIMENTS						
S-158 MULTISPECTRAL PHOTOGRAPHY		X				
S-176 APOLLO WINDOW METEOROID			X	X	X	X
S-177 UV PHOTOGRAPHY OF EARTH AND MOON				X	X	X
S-178 GEGENSCHEIN FROM LUNAR ORBIT			X	X	X	X

LUNAR ORBITAL SCIENCE EXPERIMENT ASSIGNMENTS

APOLLO 16 ORBITAL TIMELINE

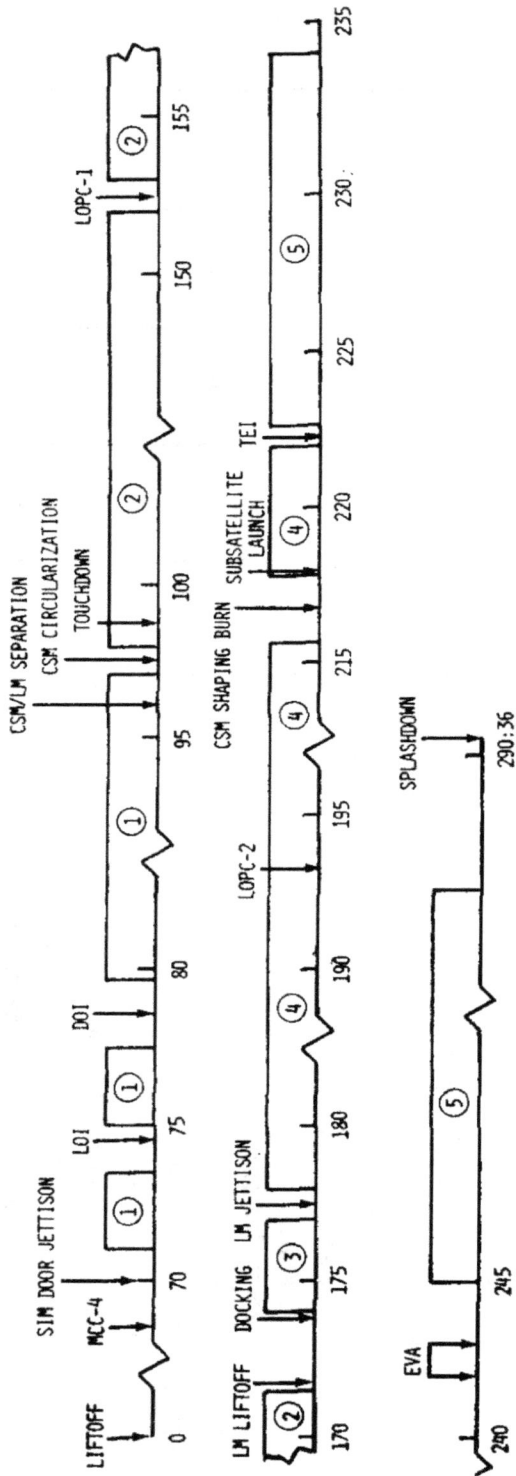

SIM OPERATION PERIODS (NON-CONTINUOUS)

ACTIVITY PERIOD	CONFIGURATION	ɣ -RAY SPEC	α-PARTICLE SPEC	X-RAY SPEC	MASS SPEC	MAP CAMERA	PAN CAMERA	LASER ALTIMETER
①	CSM-LM							
②	CSM							
③	CSM-LM							
④	CSM							
⑤	CSM							
TOTAL TIME (HOURS:MINS IN LUNAR ORBIT)		140:15	140:15	110:53	110:28	19:24	2:44	19:46

APOLLO 16
POTENTIAL LUNAR ORBIT COVERAGE

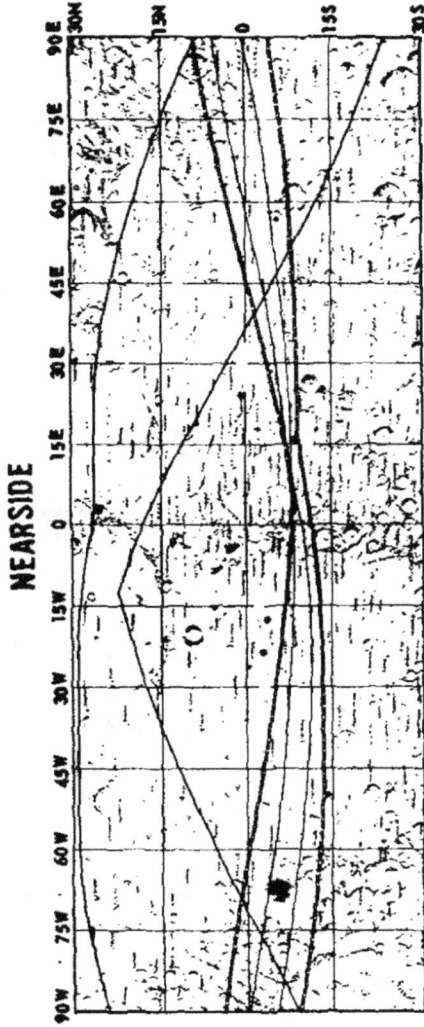

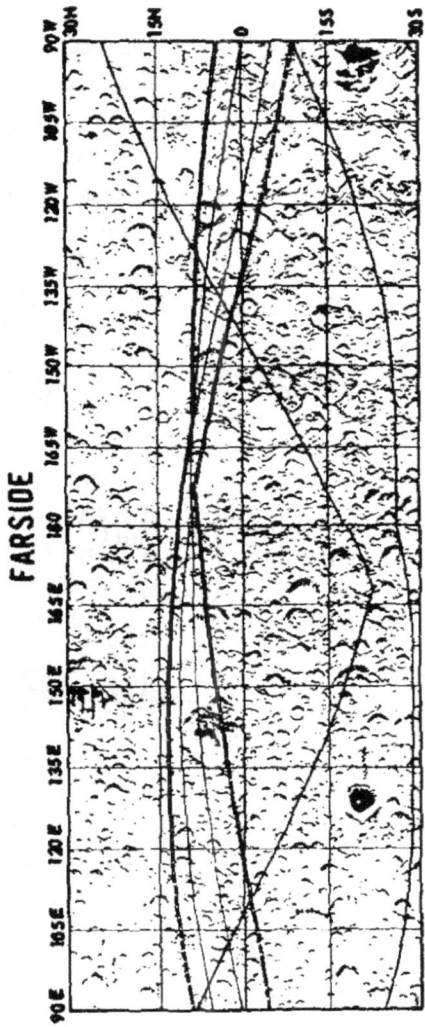

NEARSIDE

FARSIDE

LOI TO LOPC1 - 77 HR. 57 MIN. 9° INCLINATION
LOPC1 TO LOPC2 - 41 HR. 34 MIN. 10.4° INCLINATION
LOPC2 TO TEI - 28 HR. 16 MIN. 13.4° INCLINATION

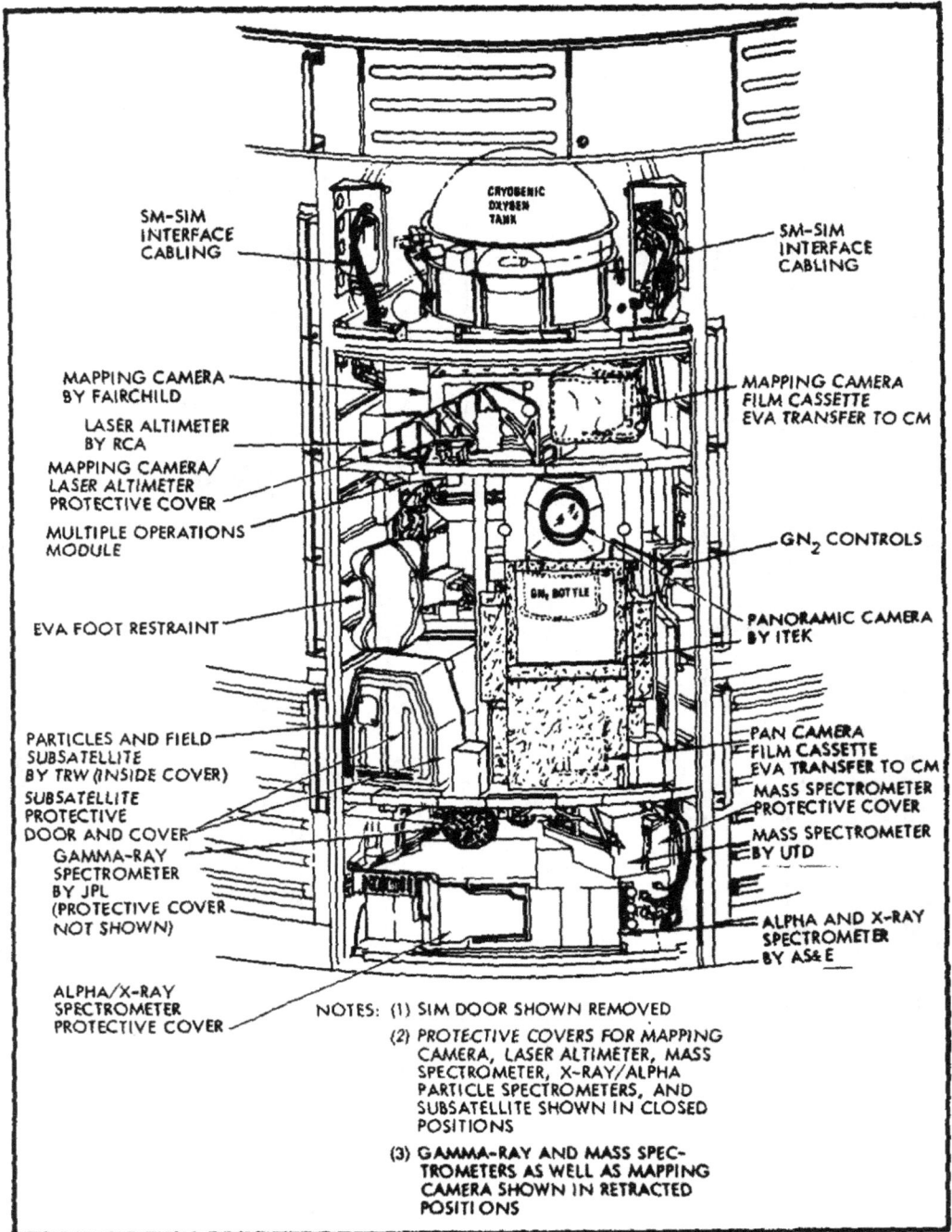

Mission SIM Bay Science Equipment Installation

Mass Spectrometer: This spectrometer measures the composition and distribution of the ambient lunar atmosphere, identi-
fies active lunar sources of volatiles, and pinpoints contamination in the lunar atmosphere. The sunset and sunrise
terminators are of special interest, since they are predicted
to be regions of concentration of certain gases. Measurements over at least five lunar revolutions are desired. The
mass spectrometer is on a 7.3-meter (24-foot) extendable boom.
The instrument can identify species from 12 to 28 atomic mass
units (AMU) with the No. 1 ion counter, and 28-66 AMU with the
No. 2 counter. Principal investigator is Dr. John H. Hoffman,
University of Texas at Dallas.

Panoramic Camera: 610mm (24-inch) SM orbital photo task:
The camera gathers mono or stereo high-resolution (2m) photographs of the lunar surface from orbit. The camera produces
an image size of 28 x 334 kilometers (17 x 208 nm) with a
field of view 11° along the track and 108° cross track. The
rotating lens system can be stowed face-inward to avoid contamination during effluent dumps and thruster firings. The
33-kilogram (72-pound) film cassette of 1,650 frames will be
retrieved by the command module pilot during a transearth
coast EVA. The camera works in conjunction with the mapping
camera and the laser altimeter to gain data to construct a
comprehensive map of the lunar surface ground track flown
by this mission ---about 2.97 million square meters (1.16
million square miles) or 8 percent of the lunar surface.

Mapping Camera 76mm (3-inch): Combines 20-meter resolution
terrain mapping photography on five-inch film with 76mm
(3-inch) focal length lens with stellar camera shooting the
star field on 35mm film simultaneously at 96° from the surface
camera optical axis. The stellar photos allow accurate orientation of mapping photography postflight by comparing simulta-
neous star field photography with lunar surface photos of
the nadir (straight down). Additionally, the stellar camera
provides pointing vectors for the laser altimeter during dark-
side passes. The mapping camera metric lens covers a 74° square
field of view, or 170 x 170 km (92 x 92 nm) from 111.5 km (60 nm)
in altitude. The stellar camera is fitted with a 76mm (3-inch)
f/2.8 lens covering a 24° field with cone flats. The 9-kg
(20-lb) film cassette containing mapping camera film (3,600
frames) and the stellar camera film will be retrieved during
the same EVA described in the panorama camera discussion.
The Apollo Orbital Science Photographic Team is headed by
Frederick J. Doyle of the U.S. Geological Survey, McLean, Va.

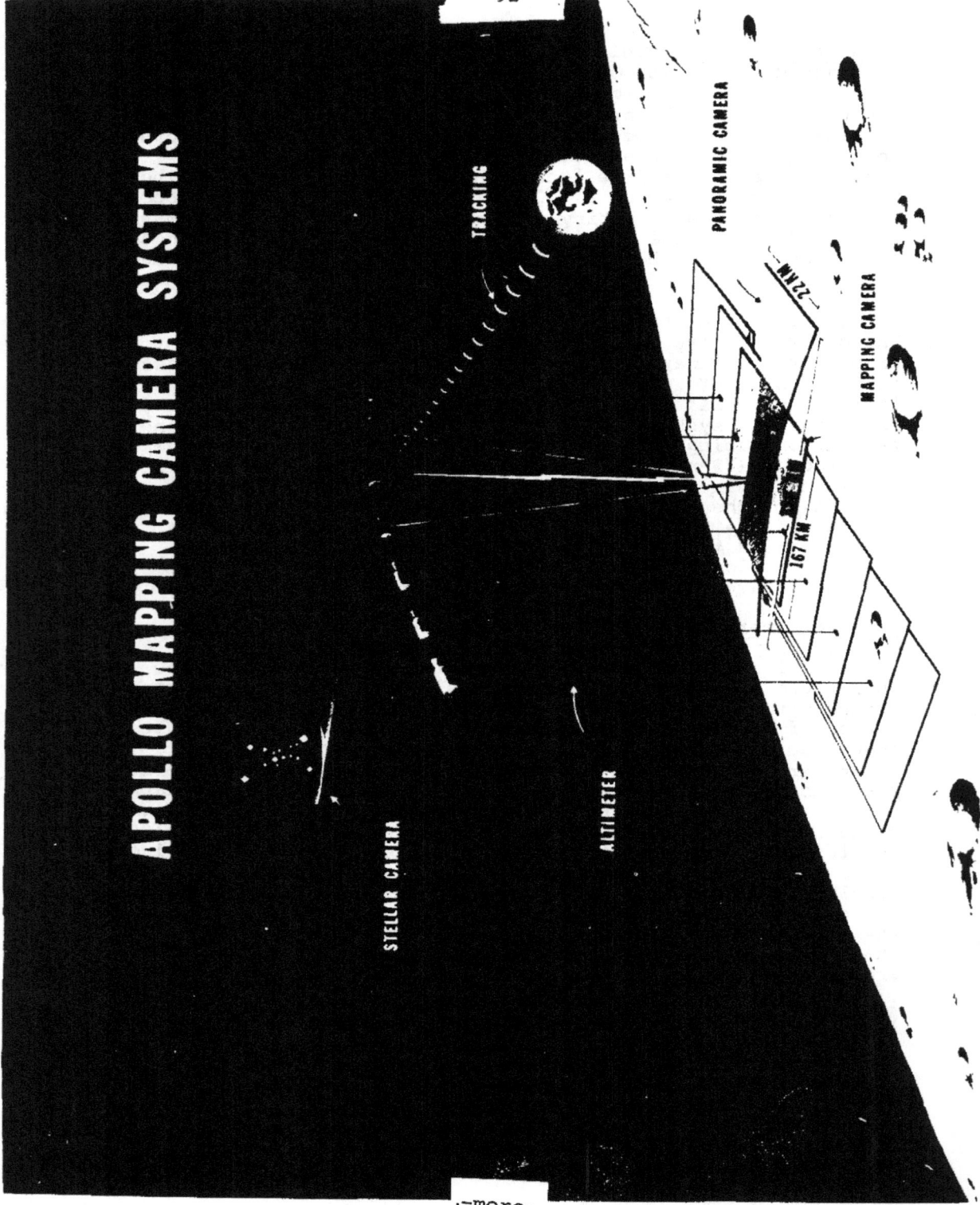

APOLLO MAPPING CAMERA SYSTEMS

STELLAR CAMERA

ALTIMETER

TRACKING

167 KM

22NM

MAPPING CAMERA

PANORAMIC CAMERA

Laser Altimeter: This altimeter measures spacecraft altitude above the lunar surface to within two meters. The instrument is boresighted with the mapping camera to provide altitude correlation data for the mapping camera as well as the panoramic camera. When the mapping camera is running, the laser altimeter automatically fires a laser pulse to the surface corresponding to mid-frame ranging for each frame. The laser light source is a pulsed ruby laser operating at 6,943 angstroms, and 200-millijoule pulses of 10 nanoseconds duration. The laser has a repetition rate up to 3.75 pulses per minute. The laser altimeter working group of the Apollo Orbital Science Photographic Team is headed by Dr. William M. Kaula of the UCLA Institute of Geophysics and Planetary Physics.

Subsatellite: The subsatellite is ejected into lunar orbit from the SIM bay and carries three experiments. The subsatellite is housed in a container resembling a rural mailbox and when deployed,is spring-ejected out-of-plane at 1.2 meters per second (4 feet per second) with a spin rate of 140 revolutions per minute. After the satellite booms are deployed, the spin rate is stabilized at about 12 rpm. The subsatellite is 77 centimeters (30 inches)long, has a 35.6 cm (14-inch hexagonal diameter and weighs 40 kg (90 pounds). The folded booms deploy to a length of 1.5 m (five feet). Subsatellite electrical power is supplied by a solar cell array outputting 24 watts for dayside operation and a rechargeable silver-cadmium battery for nightside passes.

Experiments carried aboard the subsatellite are: S-band transponder for gathering data on the lunar gravitational field, especially gravitational anomalies such as the so-called mascons particle shadows/boundary layer for gaining knowledge of the formation and dynamics of the Earth's magnetosphere, interaction of plasmas with the Moon and the physics of solar flares using telescope particle detectors and spherical electrostatic particle: detectors; and subsatellite magnetometer for gathering physical and electrical property data on the Moon and of plasma interaction with the Moon using a biaxial flux-gate magnetometer deployed on one of the three 1.5-m (5-foot) folding booms. Principal investigators for the subsatellite experiments are: particle shadows/boundary layer, Dr. Kinsey A. Anderson, University of California Berkeley; magnetometer, Dr. Paul J. Coleman, UCLA; and S-band transponder, Mr. William Sjogren, Jet Propulsion Laboratory.

-more-

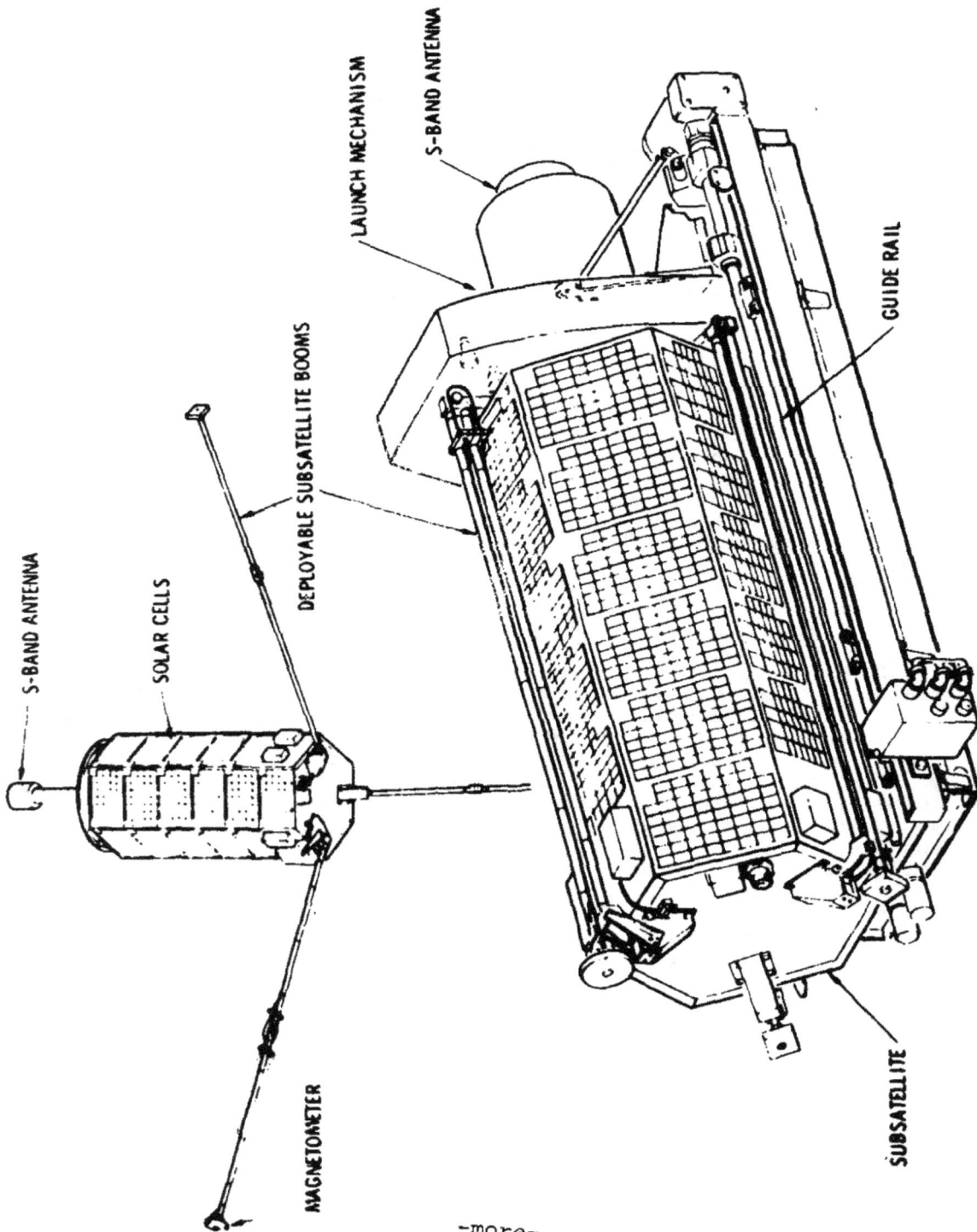

APOLLO SUBSATELLITE

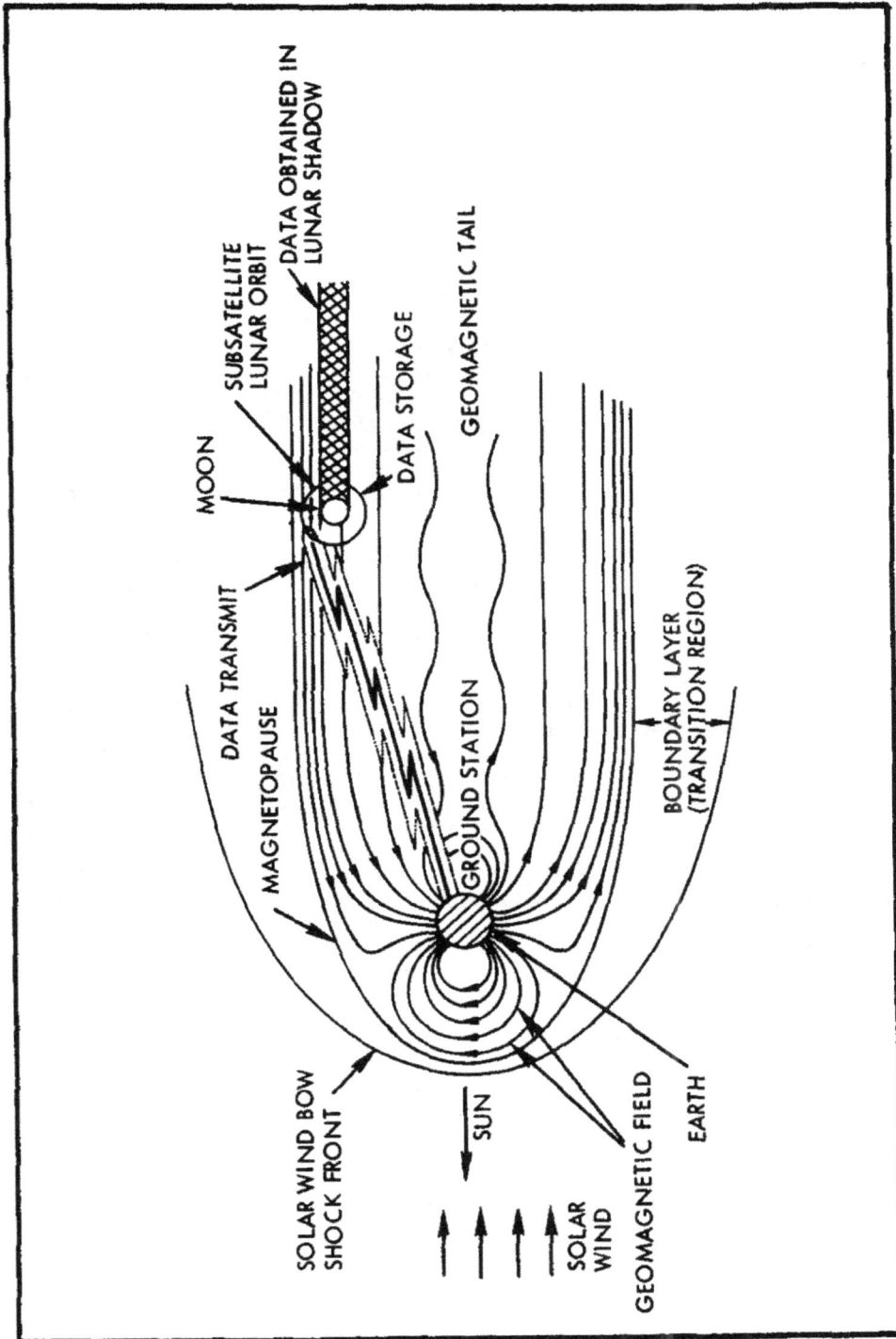

SIM Bay Subsatellite Experiments Concepts

Other CSM orbital science experiments and tasks not in SIM bay include UV photography-Earth and Moon, Gegenschein from lunar orbit, CSM/LM S-band transponder (in addition to that in the subsatellite), bistatic radar, Apollo window meteoroid experiments, microbial response, Skylab contamination, and visual flash phenomenon.

UV Photography-Earth and Moon: This experiment is aimed toward gathering ultraviolet photos of the Earth and Moon for planetary atmosphere studies and investigation of lunar surface short wavelength radiation. The photos will be made with an electric Hasselblad camera bracket-mounted in the right side window of the command module. The window is fitted with a special quartz pane that passes a large portion of the incident UV spectrum. A four-filter pack---three passing UV electromagnetic radiation and one passing visible electromagnetic radiation---is used with a 105mm lens for black and white photography; the visible spectrum filter is used with color film.

Gegenschein from Lunar Orbit: This experiment involves long exposures with a 35mm camera with 55mm f/1.2 lens on high speed black and white film (ASA 6,000). All photos must be made while the command module is in total darkness in lunar orbit.

The Gegenschein is a faint light source covering a 20° field of view along the Earth-Sun line on the opposite side of the Earth from the Sun (anti-solar axis). One theory on the origin of Gegenschein is that particles of matter are trapped at the moulton point and reflect sunlight. Moulton point is a theoretical point located 15,131,000 km (817,000 nm) from the Earth along the anti-solar axis where the sum of all gravitational forces is zero. From lunar orbit, the moulton point region can be photographed from about 15 degrees off the Earth-Sun axis, and the photos should show whether Gegenschein results from the moulton point theory or stems from zodiacal light or from some other source. The experiment was conducted on Apollo 14 and 15. The principal investigator is Lawrence Dunkelman of the Goddard Space Flight Center.

CSM/LM S-Band Transponder: The objective of this experiment is to detect variations in lunar gravity along the
lunar surface track. These anomalies in gravity result in
minute perturbations of the spacecraft motion and are
indicative of magnitude and location of mass concentrations
on the Moon. The Spaceflight Tracking and Data Network(STDN)
and the Deep Space Network (DSN) will obtain and record
S-band doppler tracking measurements from the docked
CSM/LM and the undocked CSM while in lunar orbit; S-band
doppler tracking measurements of the LM during non-powered
portions of the lunar descent; and S-band doppler tracking measurements of the LM ascent stage during non-powered
portions of the descent for lunar impact. The CSM and LM
S-band transponders will be operated during the experiment
period. The experiment was conducted on Apollo 14 and 15.

S-band doppler tracking data have been analyzed from
the Lunar Orbiter missions and definite gravity variations
were detected. These results showed the existence of mass
concentrations (mascons) in the ringed maria. Confirmation
of these results has been obtained with Apollo tracking
data.

With appropriate spacecraft orbital geometry much
more scientific information can be gathered on the lunar
gravitational field. The CSM and/or LM in low-altitude or-
bits can provide new detailed information on local gravity
anomalies. These data can also be used in conjunction with
high-altitude data to possibly provide some description on
the size and shape of the perturbing masses. Correlation
of these data with photographic and other scientific records
will give a more complete picture of the lunar environment
and support future lunar activities. Inclusion of these
results is pertinent to any theory of the origin of the
Moon and the study of the lunar subsurface structure.
There is also the additional benefit of obtaining better
navigational capabilities for future lunar missions in
that an improved lunar gravity model will be known.
William Sjogren, Jet Propulsion Laboratory, Pasadena,
California is principal investigator.

Bistatic Radar Experiment (BRE): The downlink bistatic radar
experiment seeks to measure the electromagnetic properties of
the lunar surface by monitoring that portion of the spacecraft telemetry and communications beacons which are reflected
from the Moon.

The CSM S-band telemetry beacon (f = 2.2875 gigahertz),
the VHF voice communications link (f = 259.7 megahertz), and
the spacecraft omni-directional and high gain antennas are
used in the experiment. The spacecraft is oriented so that
the radio beacon is incident on the lunar surface and is
successively reoriented so that the angle at which the signal
intersects the lunar surface is varied. The radio signal
is reflected from the surface and is monitored on Earth.
The strength of the reflected signal will vary as the angle
at which it intersects the surface is varied.

The electromagnetic properties of the surface can be
determined by measuring the reflected signal strength as a
function of angle of incidence on the lunar surface. The
angle at which the reflected signal strength is a minimum is
known as the Brewster angle and determines the dielectric
constant. The reflected signals can also be analyzed for data
on lunar surface roughness and surface electrical conductivity.

The S-band signal will primarily provide data on the
surface. However, the VHF signal is expected to penetrate the
gardened debris layer (regolith) of the Moon and be reflected
from the underlying rock strata. The reflected VHF signal
will then provide information on the depth of the regolith over
the Moon.

The S-band BRE signal will be monitored by the 210-foot
antenna at the Goldstone, California site and the VHF portion
of the BRE signal will be monitored by the 150-foot antenna
at the Stanford Research Institute in California. The experiment was flown on Apollos 14 and 15.

Lunar bistatic radar experiments were also performed
using the telemetry beacons from the unmanned Lunar Orbiter I
in 1966 and from Explorer 35 in 1967. Taylor Howard, Stanford
University, is the principal investigator.

Apollo Window Meteoroid: This is a passive experiment in which
command module windows are scanned under high magnification pre-
and postflight for evidence of meteoroid cratering flux of one-
trillionth gram or larger. Such particle flux may be a factor
in degradation of surfaces exposed to space environment.
Principal investigator is Burton Cour-Palais, NASA Manned
Spacecraft Center.

MEDICAL EXPERIMENTS AND TESTS

Apollo 16 will carry four medical experiments and tests, two of which are passive and require no crew manipulation and two in which the crew participates. The experiments are: microbial response in space environment, visual light flash phenomenon, biostack and bone mineral measurement.

Microbial Response in Space Environment: A rectangular container with some 60 million microbial passengers aboard will be attached to the command module hatch camera boom for a 10-minute period during the CMP's transearth EVA. The experiment will measure the effects of reduced oxygen pressure, vacuum, zero-g and solar ultraviolet irradiation upon five strains of bacteria, fungi and viruses. Specific strains chosen for the experiment are Rhodotorula rubra, bacillus thuringiensis, bacillus subtilis, aeromanas proteolytica and chaetomium globosum. None of these bacteria are harmful to man.

The 11.6x11.6x24.8 cm (4.5x4.5x9.75 inch) container---microbial ecology evaluation device (MEED)---contains three separate trays, each with 280 chambers for temperature sensors, ultraviolet measuring solutions, and the microorganisms. About two-thirds of the microorganisms will be in a dry state while the remainder will be in water suspension. The MEED will be opened and pointed toward the Sun for the 10-minute test period, then capped and returned to Earth for analysis.

Microbial experiments run on Geminis 9 and 12, Biosatellite 2 and in the Soviet Union's Vostok spacecraft suggest that perhaps zero-g coupled with a reduced partial pressure of oxygen has an effect upon growth and mutation rates of microorganisms.

Experiment principal investigators are: Dr. Paul Volz, Eastern Michigan University, Ypsilanti, Mich.; Dr. Bill G. Foster, Texas A&M University, College Station, Tex.; and Dr. John Spizien, Scripps Clinic and Research Foundation, La Jolla, Calif.

Visual Light Flash Phenomenon: Mysterious specks of light penetrating closed eyelids have been reported by crewmen of every Apollo lunar mission since Apollo 11. Usually the light specks and streaks are observed in a darkened command module cabin while the crew is in a rest period. Averaging two flashes a minute, the phenomena was observed in previous missions in translunar and transearth coast and in lunar orbit.

-more-

Two theories have been proposed on the origin of the flashes. One theory is that the flashes stem from visual phosphenes induced by cosmic rays. The other theory is that Cerenkov radiation by high-energy atomic particles either enter the eyeball or ionize upon collision with the retina or cerebral cortex.

The Apollo 16 crew will run a controlled experiment during translunar and transearth coast in an effort to correlate light flashes to incident primary cosmic rays. One crewman will wear an emulsion plate device on his head called the Apollo light flash moving emulsion detector (ALFMED), while his crewmates wear eye shields. The ALFMED emulsion plates cover the front and sides of the wearer's head and will provide data on time, strength, and path of high-energy atomic particles penetrating the emulsion plates. This data will be correlated with the crewman's verbal reports on flash observations during the tests.

Dr. R. E. Benson of the NASA Manned Spacecraft Center Preventive Medicine Division is the principal investigator.

Biostack: The German biostack experiment is a passive experiment requiring no crew action. Selected biological material will be exposed to high-energy heavy ions in cosmic radiation and the effects analyzed postflight. Heavy ion energy measurements cannot be gathered from ground-based radiation sources.

The experiment results will add to the knowledge of how these heavy ions may present a hazard for man during long space flights.

Alternate layers of biological materials and radiation track detectors are hermetically sealed in a cylindrical aluminum container 12.5 cm in diameter by 9.8 cm high (3.4 x 4.8 inches) and weighing 2.4 kg (5.3 lbs). The biostack container will be mounted in the command module in a position to minimize shielding to cosmic radiation. This container has no organisms harmful to man.

After command module recovery, the container will be returned to the principal investigator, Dr. Horst Bucker, University of Frankfurt, Frankfurt am Main, Federal Republic of Germany, whose participation is funded and sponsored by the German Ministry for Education and Science. Dr. Bucker will be assisted in his investigations by scientists at the University of Strasbourg, France, and at the Aerospace Medicine Research Center in Paris. The experiment is endorsed by the Working Party on Space Biophysics of the Council of Europe's parliamentary Committee on Science and Technology.

-more-

Bone Mineral Measurement: Mineral changes in human bones
caused by reduced gravity are measured in this passive experiment.
X-ray absorption techniques pre- and post-flight will measure
bone mineral content of the radius, ulna, and os calcis (heel) for
comparisons.

Principal investigator is Dr. J. M. Vogel, U.S. Public Health
Service Hospital, San Francisco, Calif.

ENGINEERING/OPERATIONAL TESTS AND DEMONSTRATION

Four tests aimed at gathering data for improving operations in future space flight missions will be flown aboard Apollo 16. An investigation into external spacecraft contamination and an evaluation of the Skylab food packaging system will provide data useful to next year's Skylab series of long-duration missions. Tests of an improved gas/water separator for removing hydrogen gas bubbles from potable water generated as a fuel cell byproduct and evaluation of a new type of fecal collection bag also will be conducted. An electrophoresis separation demonstration will also be conducted.

Skylab Contamination Study: Since John Glenn reported seeing "fireflies" outside the tiny window of his Mercury spacecraft Friendship 7 a decade ago, space crews have noted light-scattering particles that hinder visual observations as well as photographic tasks. Opinions on the origin of the clouds of particles surrounding spacecraft have ranged from ice crystals generated by water vapor dumped from the spacecraft to a natural phenomenon of particles inhabiting the space environment.

The phenomenon could be of concern in the Skylab missions during operation of the solar astronomy experiments. The Apollo 16 tests will attempt to further identify the sources of contamination outside the spacecraft. The crew will use a combination of photography and visual observations to gather the data, which in turn will serve as baselines for predicting contamination in the vicinity of the Skylab space station and for devising methods for minimizing contaminate levels around the Skylab vehicle.

Skylab Food Package: Since the Skylab orbital workshop must carry enough food to last nine men for 140 days, the methods of packaging the food must be somewhat different than the methods used in packaging food for the relatively short Apollo missions.

Packed in with the Apollo 16 crew's regular fare of meals will be several Skylab food items for the crew to evaluate. Among the types of packaging to be checked are two sizes of snap-top cans containing foods ranging from dried peaches and puddings to peanuts, wet-pack spoon-bowl foods, postage-stamp size salt dispensers, and plastic-bellows drink containers.

- more -

The Apollo 16 crew will take still and motion pictures of the food packages in use and make subjective comments on ease of handling and preparation of the food.

Improved Gas/Water Separator: During transearth coast the Apollo 16 crew will unstow an improved gas/water separator and use it on the food preparation water spigot for one meal period. Command module potable water is a byproduct of the fuel cell powerplants and contains hydrogen gas bubbles which cause crew discomfort when ingested. Gas/water separators flown previously have had leaks around the seals and have not adequately separated gas bubbles from the water.

The improved separator is expected to deliver bubble-free drinking and food preparation water.

Improved Fecal Collection Bag: Defecation bags used in previous Apollo missions are virtually identical to the Gemini "blue bags". During previous missions, the crews have experienced some difficulty in attaching the bag to the body, in post-defecation sealing, and personal hygiene clean-up. The improved fecal collection bag has been re-designed for better fit and sealing.

Electrophoretic Separation Demonstration: The Apollo 16 astronauts will use the natural weightless environment in space flight to again demonstrate an improved process in materials purification which may permit separation of materials of higher purity than can now be produced on Earth.

The space demonstration is called electrophoretic separation, referring to the movement of electrically charged particles under the influence of an electric field.

Most organic molecules become electrically charged when they are placed in a solution and will move through such a solution if an electric field is applied. Since molecules of different size and shape move at different speeds, the faster molecules in a mixture will "outrun" the slower ones as they move from one end of the solution to the other.

Thus, like particles separate into bands as they drift towards the attracting electrode. Separation is accomplished by removing the desired bands of particles.

- more -

On Earth the purification or separation process is generally performed with filters or other supporting devices because materials of any significant size rapidly sink or settle in the solution before separation occurs. Additionally, light materials remaining are easily mixed by convective currents, thus reducing purity.

The Apollo 14 electrophoretic separation demonstration experienced some technical problems which prohibited obtaining information on two of the three samples. Results with the third sample (a mixture of red and blue dyes) indicated the feasibility of obtaining sharper separations in space than on Earth. The Apollo 16 demonstration has been modified to overcome the problems encountered with the Apollo 14 demonstration and will use polystyrene particles to simulate separation of large biological particles. The Apollo 16 demonstration will allow a more accurate evaluation of the separation process.

If successful, the demonstration could show that one of the practical uses for future manned space stations may be to economically produce vaccine and medical preparations of very high purity.

LUNAR ROVING VEHICLE

The lunar roving vehicle (LRV) will transport two astronauts on three exploration traverses of the Moon's Descartes region during the Apollo 16 mission. The LRV will also carry tools, scientific and communications equipment, and lunar samples.

The four-wheel, lightweight vehicle greatly extends the lunar area that can be explored by man. It is the first manned surface transportation system designed to operate on the Moon, and it represents a solution to challenging new problems without precedent in Earth-bound vehicle design and operation.

The LRV must be folded into a small package within a wedge-shaped storage bay of the lunar module descent stage for transport to the Moon. After landing, the vehicle must be unfolded from its stowed position and deployed on the surface. It must then operate in an almost total vacuum under extremes of surface temperatures, low gravity, and on unfamiliar terrain.

The first lunar roving vehicle, used on the Apollo 15 lunar mission, was driven for three hours during its exploration traverses, covering a distance of 27.8 kilometers (17.3 statute miles) at an average speed of 9.2 kilometers an hour (5.7 miles an hour).

General Description

The LRV is 3.1 meters long (10.2 feet); has a 1.8-meter (six-foot) tread width; is 1.14 meters high (44.8 inches); and has a 2.3-meter wheel base (7.5 feet). Each wheel is powered by a small electric motor. The maximum speed reached on the Apollo 15 mission was about 13 km/hr (eight mph).

Two 36-volt batteries provide vehicle power, and either battery can run all systems. The front and rear wheels have separate steering systems; if one fails it can be disconnected and the LRV will operate with the other system.

- more -

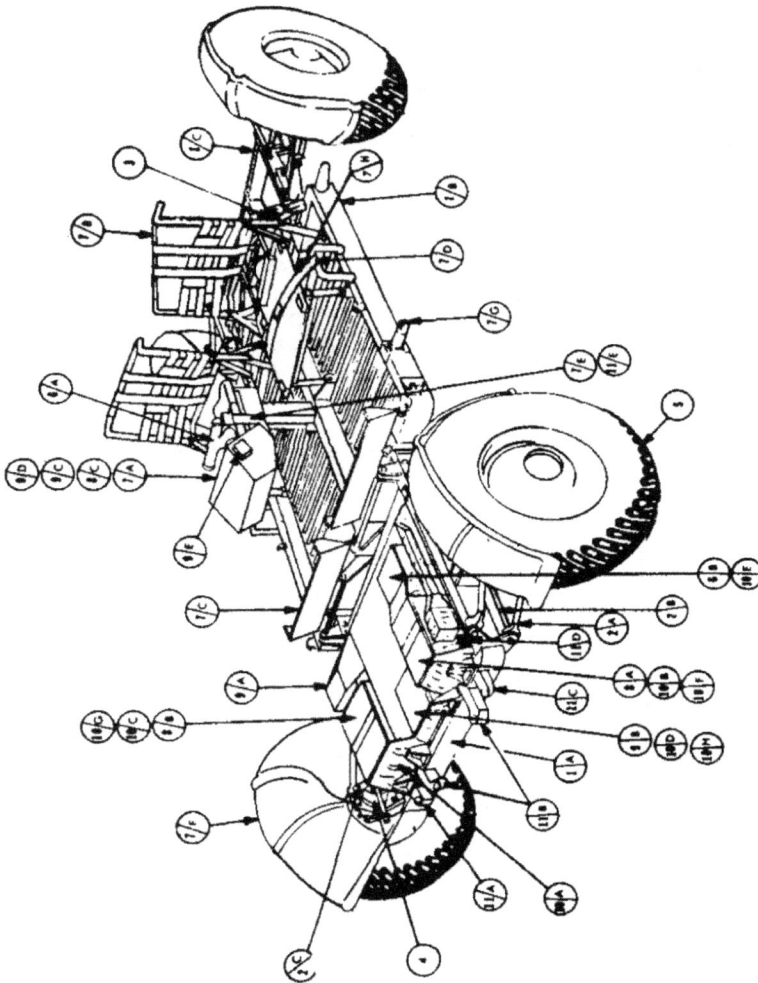

① CHASSIS

  A. FORWARD CHASSIS
  B. CENTER CHASSIS
  C. AFT CHASSIS

② SUSPENSION SYSTEM

  A. SUSPENSION ARMS (UPPER AND LOWER)
  B. TORSION BARS (UPPER AND LOWER)
  C. DAMPER

③ STEERING SYSTEM (FORWARD AND AFT)

④ TRACTION DRIVE

⑤ WHEEL

⑥ DRIVE CONTROL

  A. HAND CONTROLLER
  B. DRIVE CONTROL ELECTRONICS (DCE)

⑦ CREW STATION

  A. CONTROL AND DISPLAY CONSOLE
  B. SEAT
  C. FOOTREST
  D. OUTBOARD HANDHOLD
  E. INBOARD HANDHOLD
  F. FENDER
  G. TOEHOLD
  H. SEAT BELT

⑧ POWER SYSTEM

  A. BATTERY #1
  B. BATTERY #2
  C. INSTRUMENTATION

⑨ NAVIGATION

  A. DIRECTIONAL GYRO UNIT (DGU)
  B. SIGNAL PROCESSING UNIT (SPU)
  C. INTEGRATED POSITION INDICATOR (IPI)
  D. SUN SHADOW DEVICE
  E. VEHICLE ATTITUDE INDICATOR

⑩ THERMAL CONTROL

  A. INSULATION BLANKET
  B. BATTERY NO. 1 DUST COVER
  C. BATTERY NO. 2 DUST COVER
  D. SPU DUST COVER
  E. DCE THERMAL CONTROL UNIT
  F. BATTERY NO. 1 RADIATOR
  G. BATTERY NO. 2 RADIATOR
  H. SPU THERMAL CONTROL UNIT

⑪ PAYLOAD INTERFACE

  A. TV CAMERA RECEPTACLE
  B. LCRU RECEPTACLE
  C. HIGH GAIN ANTENNA RECEPTACLE
  D. AUXILIARY CONNECTOR
  E. LOW GAIN ANTENNA RECEPTACLE

LRV WITHOUT STOWED PAYLOAD

LRV Y = 0

- Z = 100.0 (BOTTOM OF CHASSIS)

STA. X = 116.5

24" - 970 LB. PAYLOAD
27" - NO LOAD

90"

122"

STA. X = 26.5

LRV COMPONENTS AND DIMENSIONS

44.8" MAX

14" - LOADED
17" - NO LOAD

72"

AXIS REFERENCE

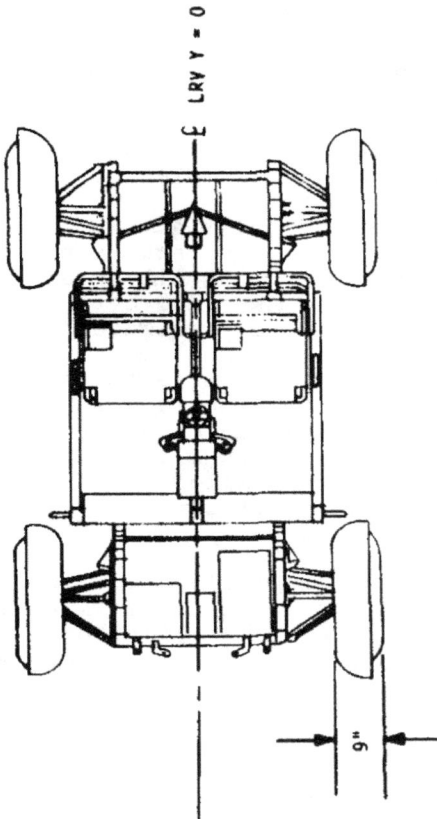

Z

+Y

X (FWD)

-Y

(DEPLOYED, EMPTY)

WEIGHT = 462 LB*

C.G. LOCATION:

X = 52.8
Y = -0.3
Z = 103.1

*INCLUDES BATTERIES & PAYLOAD SUPPORTS, EXCLUDES SSE.

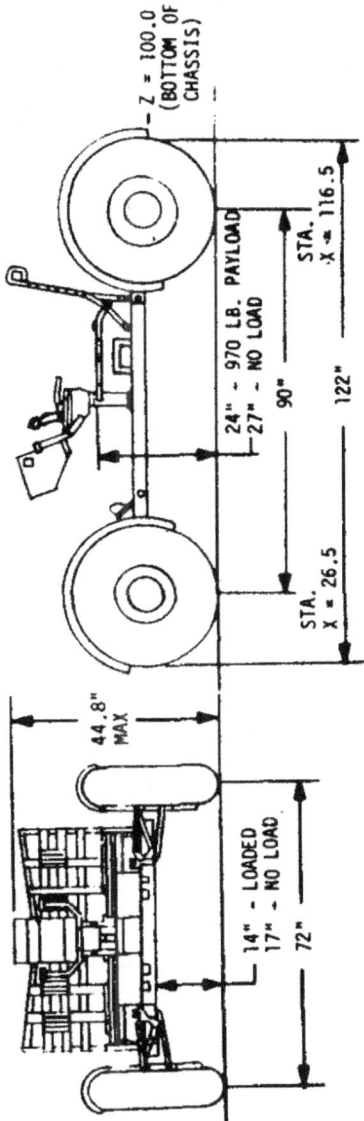

9"

-more-

Weighing about 208 kilograms (457 pounds), Earth weight when deployed on the Moon, the LRV can carry a total payload of about 490 kilograms (1,080 pounds), more than twice its own weight. The payload includes two astronauts and their portable life support systems (about 363 kilograms; 800 pounds), 45.4 kilograms (100 pounds) of communications equipment, 54.5 kilograms (120 pounds) of scientific equipment and photographic gear, and 27.2 kilograms (60 pounds) of lunar samples.

The LRV is designed to operate for a minimum of 78 hours during the lunar day. It can make several exploration sorties to a cumulative distance of 92 kilometers (57 miles). The maximum distance the LRV will be permitted to range from the lunar module will be approximately 9.7 kilometers (six miles), the distance the crew could safely walk back to the LM in the unlikely event of a total LRV failure. This walkback distance limitation is based upon the quantity of oxygen and coolant available in the astronauts' portable life support systems. This area contains about 292 square kilometers (113 square miles) available for investigation, 10 times the area that can be explored on foot.

The vehicle can negotiate obstacles 30.5 centimeters (one foot) high, and cross crevasses 70 centimeters wide (28 inches). The fully loaded vehicle can climb and descend slopes as steep as 25 degrees, and park on slopes up to 35 degrees. Pitch and roll stability angles are at least 45 degrees, and the turn radius is three meters (10 feet).

Both crewmen sit so the front wheels are visible during normal driving. The driver uses an on-board dead reckoning navigation system to determine direction and distance from the lunar module, and total distance traveled at any point during a traverse.

The LRV has five major systems: mobility, crew station, navigation, power, and thermal control. Secondary systems include the deployment mechanism, LM attachment equipment, and ground support equipment.

- more -

The aluminum chassis is divided into three sections that support all equipment and systems. The forward and aft sections fold over the center one for stowage in the LM. The forward section holds both batteries, part of the navigation system, and electronics gear for the traction drive and steering systems. The center section holds the crew station with its two seats, control and display console, and hand controller. The floor of beaded aluminum panels can support the weight of both astronauts standing in lunar gravity. The aft section holds the scientific payload.

Auxiliary LRV equipment includes the lunar communications relay unit (LCRU) and its high and low gain antennas for direct communications with Earth, the ground commanded television assembly (GCTA), a motion picture camera, scientific equipment, tools, and sample stowage bags.

Mobility System

The mobility system is the major LRV system, containing the wheels, traction drive, suspension, steering, and drive control electronics subsystems.

The vehicle is driven by a T-shaped hand controller located on the control and display console post between the crewmen. Using the controller, the astronaut maneuvers the LRV forward, reverse, left and right.

Each LRV wheel has a spun aluminum hub and a titanium bump stop (inner frame) inside the tire (outer frame). The tire is made of a woven mesh of zinc-coated piano wire to which titanium treads are riveted in a chevron pattern around the outer circumference. The bump stop prevents excessive inflection of the mesh tire during heavy impact. Each wheel weighs 5.4 kilograms (12 pounds) on Earth and is designed to be driven at least 180 kilometers (112 miles). The wheels are 81.3 centimeters (32 inches) in diameter and 22.9 centimeters (nine inches) wide.

A traction drive attached to each wheel has a motor harmonic drive gear unit, and a brake assembly. The harmonic drive reduces motor speed at an 80-to-1 rate for continuous operation at all speeds without gear shifting. The drive has an odometer pickup (measuring distance traveled) that sends data to the navigation system. Each motor develops 0.18 kilowatt (1/4-horsepower) and operates from a 36-volt input.

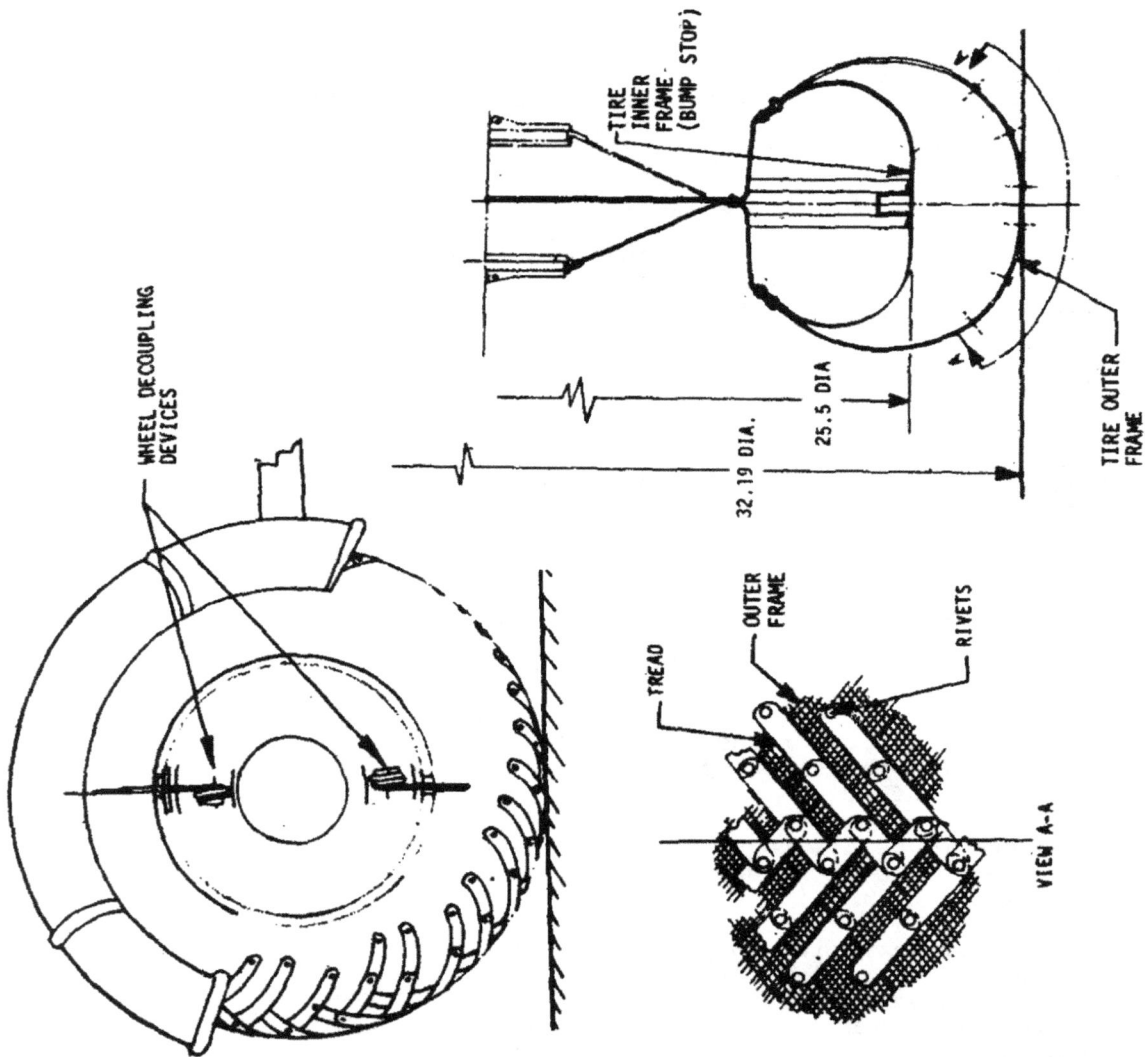

WHEEL DECOUPLING DEVICES

TIRE INNER FRAME (BUMP STOP)

TIRE OUTER FRAME

32.19 DIA.

25.5 DIA

LRV WHEEL

TREAD

OUTER FRAME

RIVETS

VIEW A-A

Each wheel has a mechanical brake connected to the hand
controller. Moving the controller rearward de-energizes
the drive motors and forces brake shoes against a drum,
stopping wheel hub rotation. Full rear movement of the
controller engages and locks a parking brake.

The chassis is suspended from each wheel by two parallel arms mounted on torsion bars and connected to each traction drive. Tire deflection allows a 35.6-centimeter
(14-inch) ground clearance when the vehicle is fully loaded,
and 43.2 centimeters (17 inches) when unloaded.

Both front and rear wheels have independent steering
systems that allow a "wall-to-wall" turning radius of
3.1 meters (122 inches), exactly the vehicle length. If
either set of wheels has a steering failure, its steering
system can be disengaged and the traverse can continue with
the active steering assembly. Each wheel can also be
manually uncoupled from the traction drive and brake to
allow "free wheeling" about the drive housing.

Pushing the hand controller forward increases forward
speed; rear movement reduces speed. Forward and reverse
are controlled by a knob on the controller's vertical stem.
With the knob pushed down, the controller can only be
pivoted forward; with it pushed up, the controller can be
pivoted to the rear for reverse.

Crew Station

The crew station consists of the control and display
console, seats, seat belts, an armrest, footrests, inboard
and outboard handholds, toeholds, floor panels, and fenders.

The control and display console is separated into two
main parts: The top portion holds navigation system displays; the lower portion contains monitors and controls.
Attached to the upper left side of the console is an
attitude indicator that shows vehicle pitch and roll.

At the console top left is a position indicator. Its
outer circumference is a large dial that shows vehicle
heading (direction) with respect to lunar north. Inside
the dial are three digital indicators that show bearing and
range to the LM and distance traveled by the LRV. In the
middle of the console upper half is a Sun compass device that
is used to update the LRV's navigation system. Down the
left side of the console lower half are control switches for
power distribution, drive and steering, and monitors for
power and temperature. A warning flag atop the console
pops up if a temperature goes above limits in either battery
or in any drive motor.

- more -

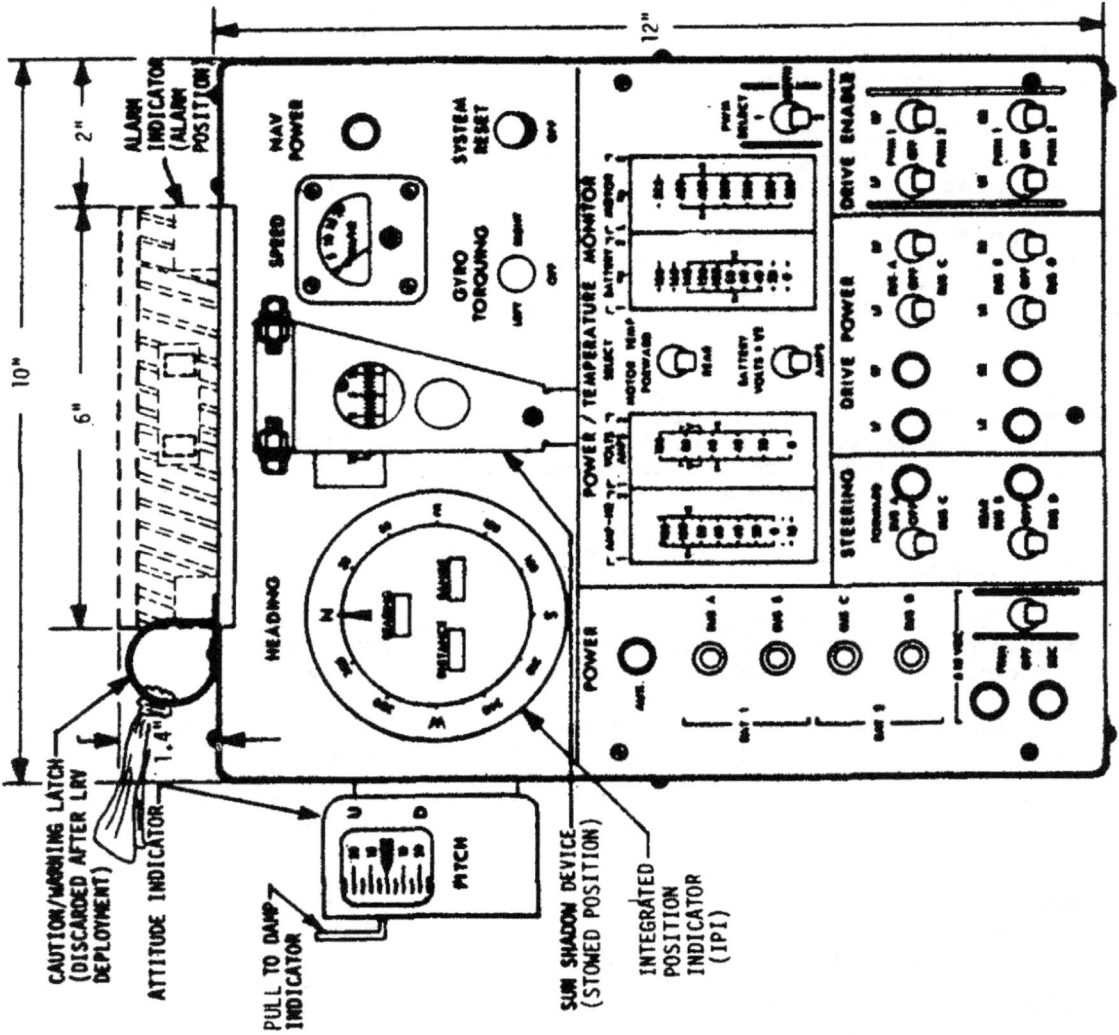

LRV CREW STATION COMPONENTS - CONTROL AND DISPLAY CONSOLE

HAND CONTROLLER OPERATION:

T-HANDLE PIVOT FORWARD - INCREASED DEFLECTION FROM NEUTRAL INCREASES FORWARD SPEED.

T-HANDLE PIVOT REARWARD - INCREASED DEFLECTION FROM NEUTRAL INCREASES REVERSE SPEED.

T-HANDLE PIVOT LEFT - INCREASED DEFLECTION FROM NEUTRAL INCREASES LEFT STEERING ANGLE.

T-HANDLE PIVOT RIGHT - INCREASED DEFLECTION FROM NEUTRAL INCREASES RIGHT STEERING ANGLE.

T-HANDLE DISPLACED REARWARD - REARWARD MOVEMENT INCREASES BRAKING FORCE. FULL 3 INCH REARWARD APPLIES PARKING BRAKE. MOVING INTO BRAKE POSITION DISABLES THROTTLE CONTROL AT 15° MOVEMENT REARWARD.

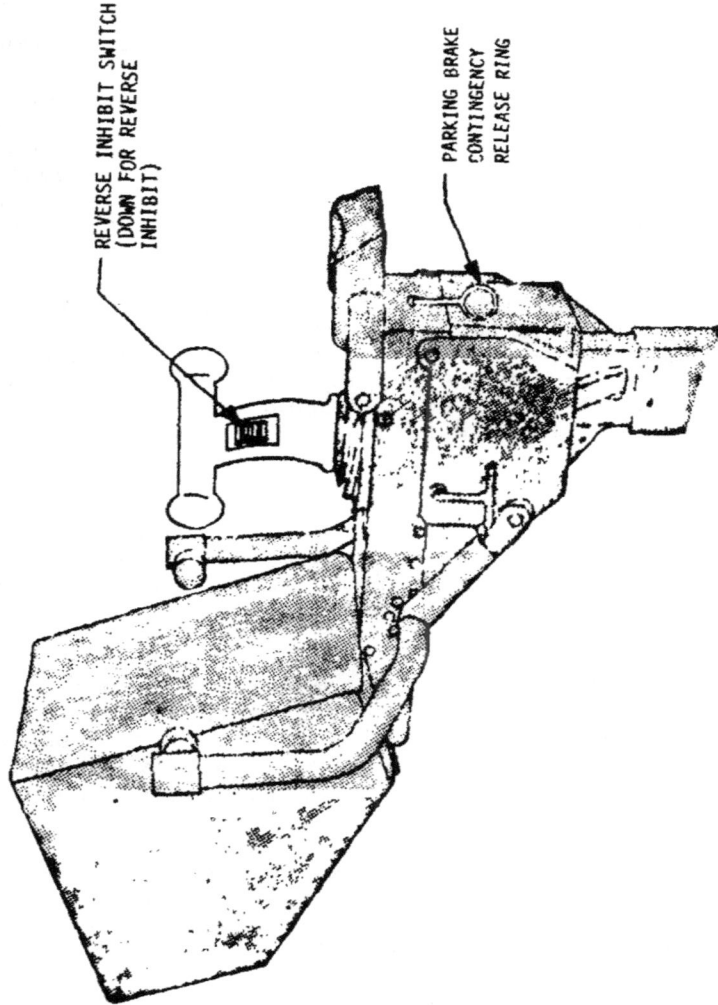

REVERSE INHIBIT SWITCH (DOWN FOR REVERSE INHIBIT)

PARKING BRAKE CONTINGENCY RELEASE RING

HAND CONTROLLER

The LRV seats are tubular aluminum frames spanned by nylon webbing. They are folded flat during launch and erected by crewmen after deployment. The seat backs support the astronaut portable life support systems. Nylon webbing seat belts, custom fitted to each crewman, snap over the outboard handholds with metal hooks.

The armrest, located directly behind the LRV hand controller, supports the arm of the driving crewman. The footrests, attached to the center floor section, are adjusted before launch to fit each crewman. Inboard handholds help crewmen get in and out of the LRV, and have receptacles for a 16 mm camera and the low gain antenna of the LCRU.

The lightweight, fiberglass fenders keep lunar dust from being thrown on the astronauts, their equipment, sensitive vehicle parts, and from obstructing vision while driving. Front and rear fender sections are retracted during flight and extended by crewmen after LRV deployment on the lunar surface.

Navigation System

The navigation system is based on the principle of starting a sortie from a known point, recording speed, direction and distance traveled, and periodically calculating vehicle position.

The system has three major components: a directional gyroscope to provide vehicle headings; odometers on each wheel's traction drive unit to give speed and distance data; and a signal processing unit (a small, solid-state computer) to determine heading, bearing, range, distance traveled, and speed.

All navigation system readings are displayed on the control console. The system is reset at the beginning of each traverse by pressing a system reset button that moves all digital displays and internal registers to zero.

The directional gyroscope is aligned by measuring the inclination of the LRV (using the attitude indicator) and measuring vehicle orientation with respect to the Sun (using the Sun compass). This information is relayed to ground controllers and the gyro is adjusted to match calculated values read back to the crew.

- more -

Each LRV wheel revolution generates odometer magnetic
pulses that are sent to the console displays.

Power System

The power system consists of two 36-volt, non-recharge-
able batteries and equipment that controls and monitors
electrical power. The batteries are in magnesium cases,
use plexiglass monoblock (common cell walls) for internal
construction, and have silver-zinc plates in potassium
hydroxide electrolyte. Each battery has 23 cells and a
121-ampere-hour capacity.

Both batteries are used simultaneously with an approximately equal load during LRV operation. Each battery can
carry the entire electrical load; if one fails, its load
can be switched to the other.

The batteries are activated when installed on the LRV
at the launch pad about five days before launch. During
LRV operation all mobility system power is turned off if a
stop exceeds five minutes, but navigation system power
remains on throughout each sortie. The batteries normally
operate at temperatures of 4.4 to 51.7 degrees C. (40-125
degrees F.).

An auxiliary connector at the LRV's forward end supplies
150 watts of 36-volt power for the lunar communications
relay unit.

Thermal Control

The basic concept of LRV thermal control is heat
storage during vehicle operation and radiation cooling when
it is parked between sorties. Heat is stored in several
thermal control units and in the batteries. Space radiators
are protected from dust during sorties by covers that are
manually opened at the end of each sortie; when battery
temperatures cool to about 7.2 degrees C. (45 degrees F.),
the covers automatically close.

- more -

A multi-layer insulation blanket protects forward
chassis components. Display console instruments are mounted
to an aluminum plate isolated by radiation shields and fiberglass mounts. Console external surfaces are coated with
thermal control paint and the face plate is anodized, as
are handholds, footrests, tubular seat sections, and center
and aft floor panels.

Stowage and Deployment

Space support equipment holds the folded LRV in the
lunar module during transit and deployment at three
attachment points with the vehicle's aft end pointing up.

Deployment is essentially manual. One crewman releases
a cable attached to the top (aft end) of the folded LRV
as the first step in the deployment. This cable is held
taut during deployment, until all four LRV wheels are on the
lunar surface.

One of the crewmen then ascends the LM ladder part way
and pulls a D-ring on the side of the descent stage. This
releases the LRV, and lets the vehicle swing out at the top
about 12.7 centimeters (five inches) until it is stopped
by two steel cables. Descending the ladder, the crewman
walks to the LRV's right side, takes the end of a deployment tape from a stowage bag, and pulls the tape hand-over-
hand. This unreels two support cables that swivel the
vehicle outward from the top. As the aft chassis is
unfolded, the aft wheels automatically unfold and deploy,
and all latches are engaged. The crewman continues to
unwind the tape, lowering the LRV's aft end to the surface,
and the forward chassis and wheels spring open and into place.

When the aft wheels are on the surface, the crewman
removes the support cables and walks to the vehicle's
left side. There he pulls a second tape that lowers the
LRV's forward end to the surface and causes telescoping
tubes to push the vehicle away from the LM. The two crewmen then deploy the fender extensions, set up the control
and display console, unfold the seats, and deploy other
equipment.

One crewman will board the LRV and make sure all
controls are working. He will back the vehicle away slightly
and drive it to the LM quadrant that holds the auxiliary
equipment. The LRV will be powered down while the crewmen
load auxiliary equipment aboard the vehicle.

- 117 -

LRV DEPLOYMENT SEQUENCE

A
- LRV STOWED IN QUADRANT
- ASTRONAUT REMOVES INSULATION BLANKET, OPERATING TAPES
- ASTRONAUT REMOTELY INITIATES DEPLOYMENT

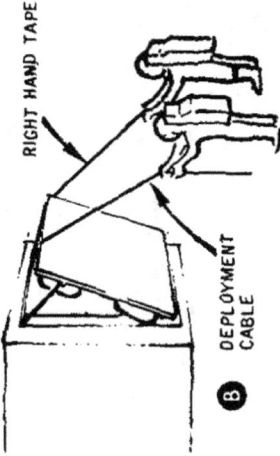

B
RIGHT HAND TAPE
DEPLOYMENT CABLE
- ASTRONAUT LOWERS LRV FROM STORAGE BAY WITH RIGHT HAND TAPE

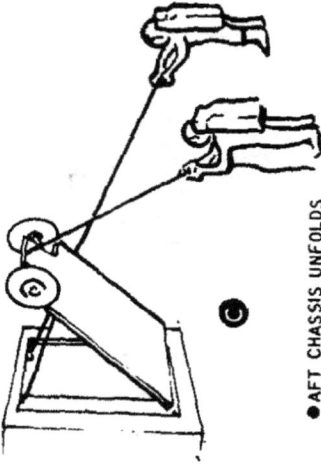

C
- AFT CHASSIS UNFOLDS
- REAR WHEELS UNFOLD
- AFT CHASSIS LOCKS IN POSITION

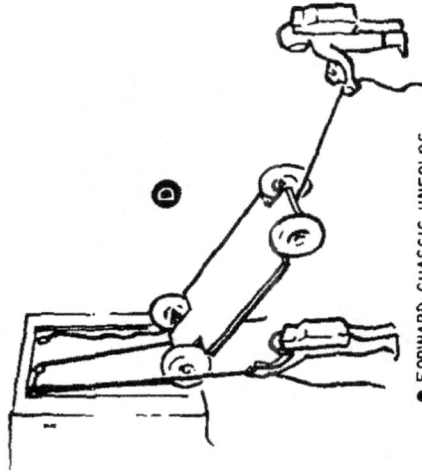

D
- FORWARD CHASSIS UNFOLDS AND LOCKS
- FRONT WHEELS UNFOLD
- ASTRONAUT LOWERS LRV TO SURFACE WITH LEFT HAND TAPE

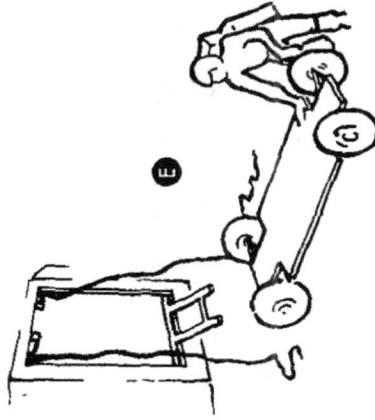

E
- ASTRONAUT DISCONNECTS SPACE SUPPORT EQUIPMENT(SSE)

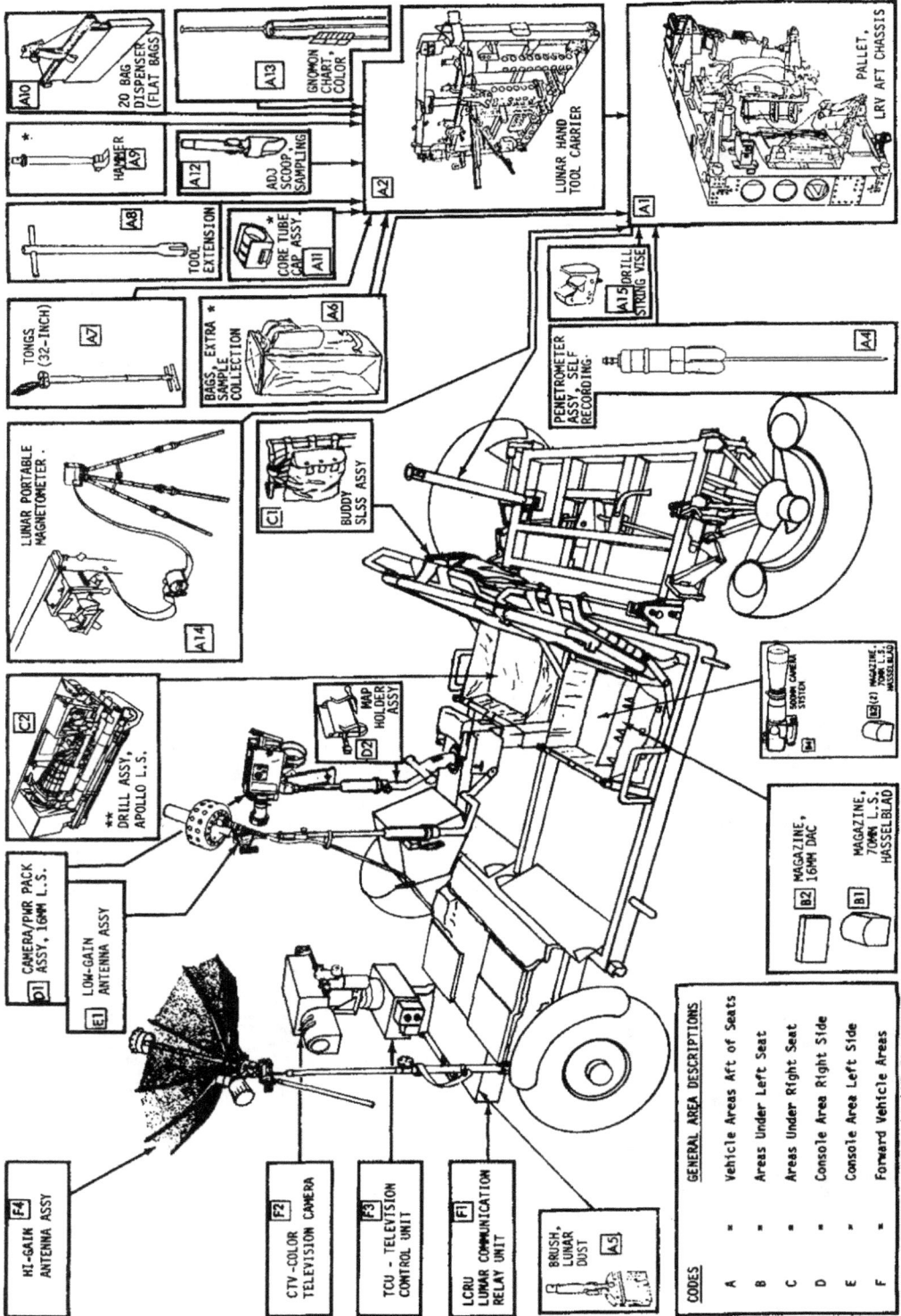

LUNAR FIELD GEOLOGY EQUIP. STOWAGE ON LRV

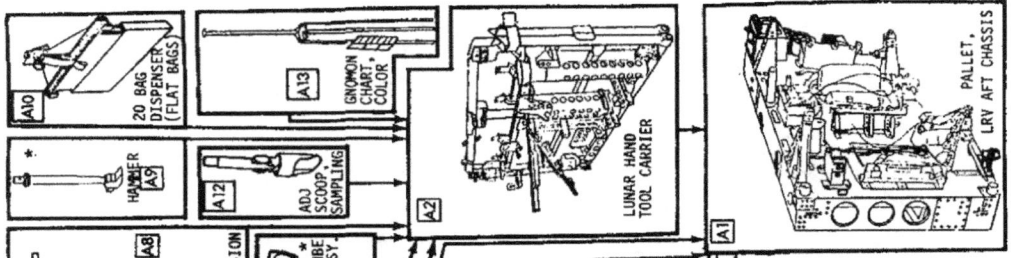

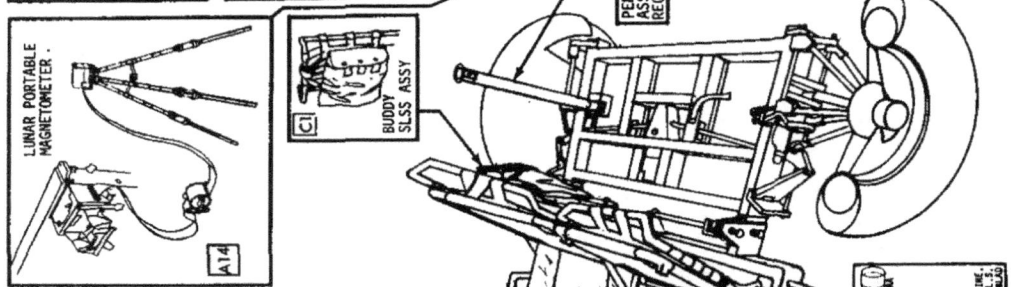

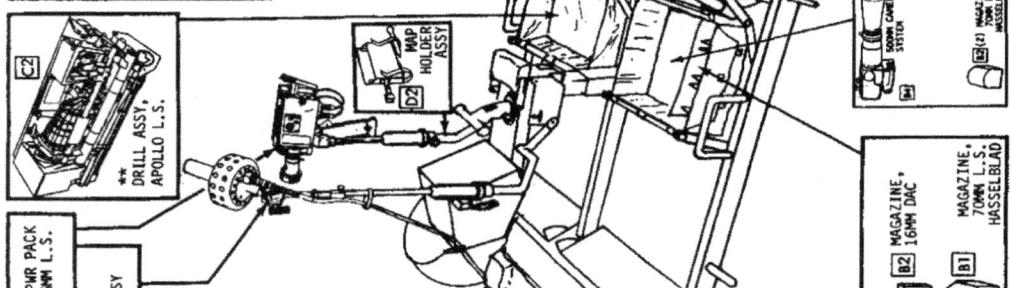

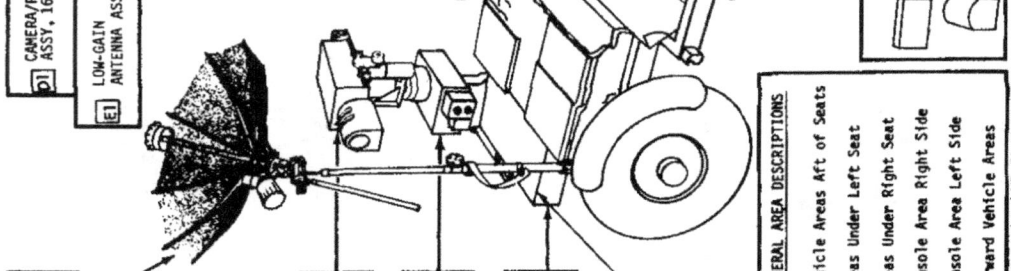

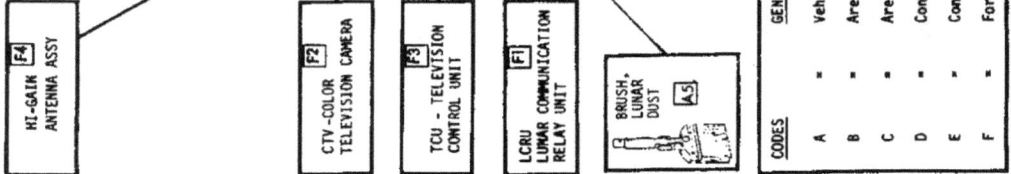

CODES		GENERAL AREA DESCRIPTIONS
A	=	Vehicle Areas Aft of Seats
B	=	Areas Under Left Seat
C	=	Areas Under Right Seat
D	=	Console Area Right Side
E	=	Console Area Left Side
F	=	Forward Vehicle Areas

*Normally carried on the LMP PLSS tool carrier
**Only carried from LM to ALSEP site

LUNAR COMMUNICATIONS RELAY UNIT (LCRU)

The range from which an Apollo crew can operate from the lunar module during EVAs while maintaining contact with the Earth is extended over the lunar horizon by a suitcase-size device called the lunar communications relay unit (LCRU). The LCRU acts as a portable relay station for voice, TV, and telemetry directly between the crew and Mission Control Center instead of through the lunar module communications system. First use of the LCRU was on Apollo 15.

Completely self-contained with its own power supply and erectable hi-gain S-Band antenna, the LCRU may be mounted on a rack at the front of the lunar roving vehicle (LRV) or hand-carried by a crewman. In addition to providing communications relay, the LCRU receives ground-command signals for the ground commanded television assembly (GCTA) for remote aiming and focusing the lunar surface color television camera. The GCTA is described in another section of this press kit.

Between stops with the lunar roving vehicle, crew voice is beamed Earthward by a wide beam-width helical S-Band antenna. At each traverse stop, the crew must boresight the high-gain parabolic antenna toward Earth before television signals can be transmitted. VHF signals from the crew portable life support system (PLSS) transceivers are converted to S-band by the LCRU for relay to the ground, and conversely, from S-Band to VHF on the uplink to the EVA crewmen.

The LCRU measures 55.9 x 40.6 x 15.2 cm (22 x 16 x 6 inches) not including antennas, and weighs 25 Earth kg (55 Earth pounds) (9.2 lunar pounds). A protective thermal blanket around the LCRU can be peeled back to vary the amount of radiation surface which consists of 1.26 m^2 (196 square inches) of radiating mirrors to reflect solar heat. Additionally, wax packages on top of the LCRU enclosure stabilize the LCRU temperature by a melt-freeze cycle. The LCRU interior is pressurized to 7.5 psia differential (one-half atmosphere).

Internal power is provided to the LCRU by a 19-cell silver-zinc battery with a potassium hydroxide electrolyte. The battery weighs 4.1 kg (nine Earth pounds) (1.5 lunar pounds) and measures 11.8 x 23.9 x 11.8 cm (4.7 x 9.4 x 4.65 inches). The battery is rated at 400 watt hours, and delivers 29 volts at a 3.1-ampere current load. The LCRU may also be operated from the LRV batteries. The nominal plan is to operate the LCRU using LRV battery power during EVA-1 and EVA-2. The LCRU battery will provide the power during EVA-3.

Three types of antennas are fitted to the LCRU system: a low-gain helical antenna for relaying voice and data when the LRV is moving and in other instances when the high-gain antenna is not deployed; a .76 m (three-foot) diameter parabolic rib-mesh high-gain antenna for relaying a television signal; and a VHF omni-antenna for receiving crew voice and data from the PLSS transceivers. The high-gain antenna has an optical sight which allows the crewman to boresight on Earth for optimum signal strength. The Earth subtends one-half degree angle when viewed from the lunar surface.

The LCRU can operate in several modes: mobile on the LRV, fixed base such as when the LRV is parked, or hand-carried in contingency situations such as LRV failure.

TELEVISION AND
GROUND COMMANDED TELEVISION ASSEMBLY

Two different color television cameras will be used during the Apollo 16 mission. One, manufactured by Westinghouse, will be used in the command module. It will be fitted with a 5 cm (2 in) black and white monitor to aid the crew in focus and exposure adjustment.

The other camera, manufactured by RCA, is for lunar surface use and will be operated from three different positions-- mounted on the LM MESA, mounted on a tripod and connected to the LM by a 30.5 m (100 ft) cable, and installed on the LRV with signal transmission through the lunar communication relay unit rather than through the LM communications system as in the other two positions.

While on the LRV, the camera will be mounted on the ground commanded television assembly (GCTA). The camera can be aimed and controlled by astronauts or it can be remotely controlled by personnel located in the Mission Control Center. Remote command capability includes camera "on" and "off", pan, tilt, zoom, iris open/closed (f2.2 to f22) and peak or average automatic light control.

The GCTA is capable of tilting the TV camera upward 85 degrees, downward 45 degrees, and panning the camera 340 degrees between mechanical stops. Pan and tilt rates are three degrees per second.

The TV lens can be zoomed from a focal length of 12.5mm to 75mm corresponding to a field of view from three to nine degrees.

At the end of the third EVA, the crew will park the LRV about 91.4 m (300 ft) east of the LM so that the color TV camera can cover the LM ascent from the lunar surface. Because of a time delay in a signal going the quarter million miles out to the Moon, Mission Control must anticipate ascent engine ignition by about two seconds with the tilt command.

The GCTA and camera each weigh approximately 5.9 kg (13 lb). The overall length of the camera is 46 cm (18.0 in) its width is 17 cm (6.7 in), and its height is 25 cm (10 in). The GCTA and LCRU are built by RCA.

-more-

APOLLO 16 TELEVISION EVENTS

DATE	TIME (GET)	TIME (EST)	DURATION (HRS:MIN)	EVENT
APRIL 16	3:05	1559	0:19	TD & E
APRIL 20	102:25	1919	6:48	EVA-1
APRIL 21	124:50	1744	6:30	EVA-2
APRIL 22	148:25	1719	6:40	EVA-3
APRIL 23	171:40	1634	0:14	LM LIFT-OFF
	173:20	1814	0:06	RENDEZVOUS
	173:44	1838	0:07	DOCKING
APRIL 26	241:55	1449	1:10	CMP EVA
TBD				TRANSEARTH COAST

PHOTOGRAPHIC EQUIPMENT

Still and motion pictures will be made of most space-craft maneuvers and crew lunar surface activities. During lunar surface operations, emphasis will be on documenting placement of lunar surface experiments, documenting lunar samples, and on recording in their natural state the lunar surface features.

Command module lunar orbit photographic tasks and experiments include high-resolution photography to support future landing missions, photography of surface features of special scientific interest and astronomical phenomena such as solar corona, Gegenschein, zodiacal light, libration points, and galactic poles.

Camera equipment stowed in the Apollo 16 command module consists of one 70mm Hasselblad electric camera, a 16mm Maurer motion picture camera, and a 35mm Nikon F single-lens reflex camera. The command module Hasselblad electric camera is normally fitted with an 80mm f/2.8 Zeiss Planar lens, but a bayonet-mount 250mm lens can be fitted for long-distance Earth/Moon photos. A 105mm f/4.3 Zeiss UV Sonnar is provided for the ultraviolet photography experiment.

The 35mm Nikon F is fitted with a 55mm f/1.2 lens for the Gegenschein and dim-light photographic experiments.

The Maurer 16mm motion picture camera in the command module has lenses of 10, 18 and 75mm focal length available. Accessories include a right-angle mirror, a power cable and a sextant adapter which allows the camera to film through the navigation sextant optical system.

Cameras stowed in the lunar module are two 70mm Hasselblad data cameras fitted with 60mm Zeiss Metric lenses, an electric Hasselblad with 500mm lens and two 16mm Maurer motion picture cameras with 10mm lenses. One of the Hasselblads and one of the motion picture cameras are stowed in the modular equipment stowage assembly (MESA) in the LM descent stage.

The LM Hasselblads have crew chest mounts that fit dovetail brackets on the crewman's remote control unit, thereby leaving both hands free. One of the LM motion picture cameras will be mounted in the right-hand window to record descent, landing, ascent and rendezvous. The 16mm camera stowed in the MESA will be carried aboard the lunar roving vehicle to record portions of the three EVAs.

- more -

- 124 -

Descriptions of the 24-inch panoramic camera and the
3-inch mapping/stellar camera are in the orbital science
section of this press kit.

- more -

TV AND PHOTOGRAPHIC EQUIPMENT

NOMENCLATURE	CSM AT LAUNCH	LM AT LAUNCH	CM TO LM	LM TO CM	CM AT ENTRY
TV, COLOR, ZOOM LENS (MONITOR WITH CM SYSTEM)	1	1			1
CAMERA, DATA ACQUISITION, 16 MM	1	1			1
LENS - 10 MM	1	1			
- 18 MM	1				1
- 75 MM	1				1
FILM MAGAZINES	13				13
CAMERA, 35 MM NIKON	1				1
LENS - 55 MM	1				1
CASSETTE, 35 MM	9				9
CAMERA, 16 MM, BATTERY OPERATED (LUNAR SURFACE)		1			
LENS - 10 MM		1			
FILM MAGAZINES	8		8	8	8
CAMERA, HASSELBLAD, 70 MM ELECTRIC	1				1
LENS - 80 MM	1				1
- 250 MM	1				1
- 105 MM UV (4 BAND-PASS FILTERS)	1				1
FILM MAGAZINES	7				7
FILM MAGAZINE, 70 MM UV	1				1
CAMERA, HASSELBLAD ELECTRIC DATA (LUNAR SURFACE)		2			
LENS - 60 MM		2			
FILM MAGAZINES	11		11	11	11
POLARIZING FILTER		1			
CAMERA, 24-IN. PANORAMIC (IN SIM)	1				
FILM MAGAZINE (EVA TRANSFER)	1				1
CAMERA, LUNAR SURFACE ELECTRIC		1			
LENS - 500 MM		1			
FILM MAGAZINES	2		2	2	2
CAMERA, 3-IN MAPPING STELLAR(SIM)	1				
FILM MAGAZINE CONTAINING 5-IN. MAPPING AND 35 MM STELLAR FILM (EVA TRANSFER)	1				1
CAMERA, ULTRAVIOLET, LUNAR SURFACE		1			
FILM MAGAZINE, UV, LS		1		1	1

ASTRONAUT EQUIPMENT

Space Suit

Apollo crewmen wear two versions of the Apollo space suit: the command module pilot version (A-7LB-CMP) for intravehicular operations in the command module and for extravehicular operations during SIM bay film retrieval during transearth coast; and the extravehicular version (A-7LB-EV) worn by the commander and lunar module pilot for lunar surface EVAs.

The A-7LB-EV suit differs from Apollo suits flown prior to Apollo 15 by having a waist joint that allows greater mobility while the suit is pressurized--stooping down for setting up lunar surface experiments, gathering samples and for sitting on the lunar roving vehicle.

From the inside out, an integrated thermal meteroid suit cover layer worn by the commander and lunar module pilot starts with rubber-coated nylon and progresses outward with layers of nonwoven Dacron, aluminized Mylar film and Beta marquisette for thermal radiation protection and thermal spacers, and finally with a layer of nonflammable Teflon-coated Beta cloth and an abrasion-resistant layer of Teflon fabric--a total of 18 layers.

Both types of the A-7LB suit have a pressure retention portion called a torso limb suit assembly consisting of neoprene coated nylon and an outer structural restraint layer.

The space suit with gloves, and dipped rubber convolutes which serve as the pressure layer liquid cooling garment, portable life support system (PLSS), oxygen purge system, lunar extravehicular visor assembly, and lunar boots make up the extravehicular mobility unit (EMU). The EMU provides an extravehicular crewman with life support for a seven-hour mission outside the lunar module without replenishing expendables.

Lunar extravehicular visor assembly - The assembly consists of polycarbonate shell and two visors with thermal control and optical coatings on them. The EVA visor is attached over the pressure helmet to provide impact, micrometeoroid, thermal and ultraviolet-infrared light protection to the EVA crewmen.

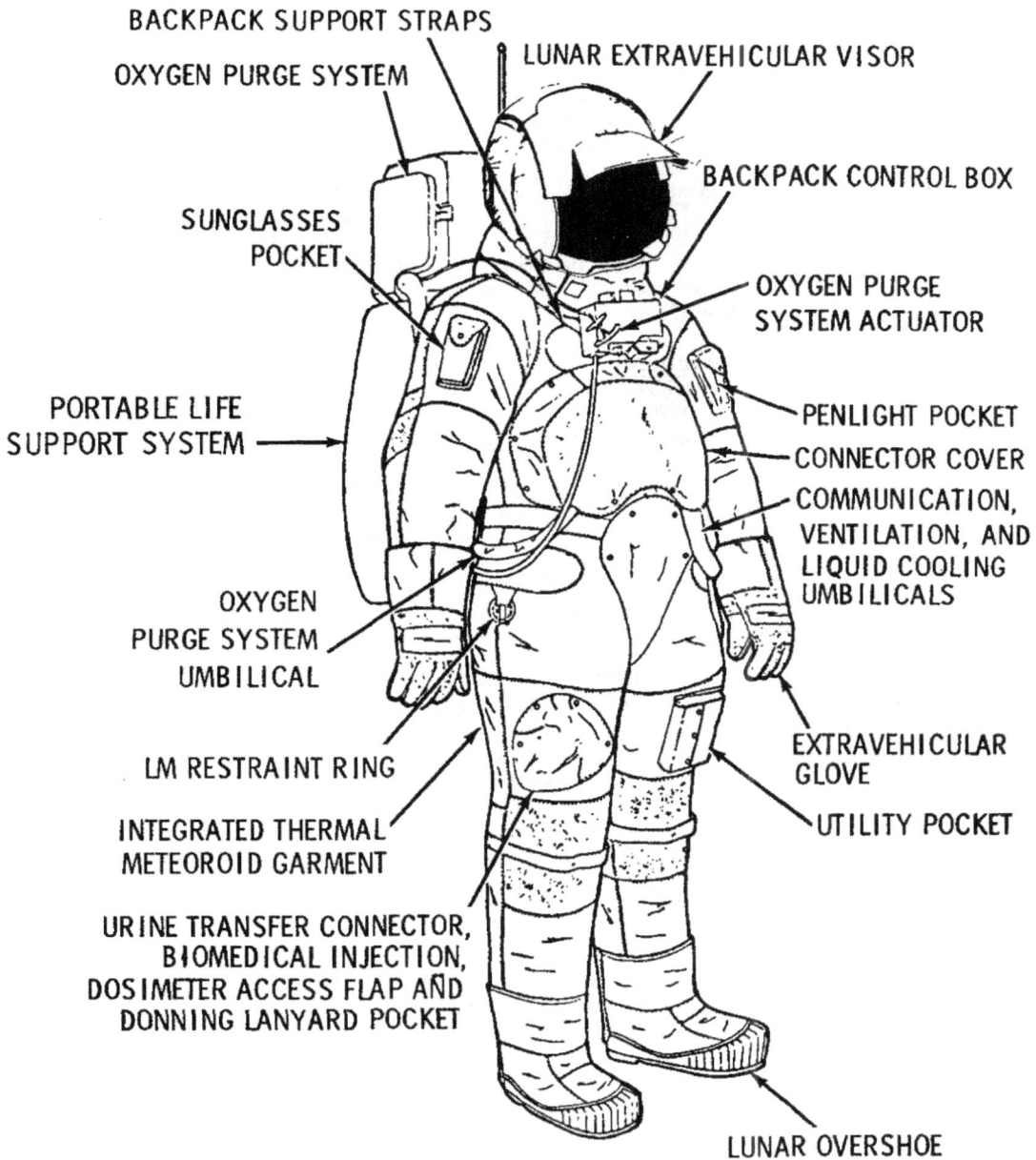

BACKPACK SUPPORT STRAPS

OXYGEN PURGE SYSTEM

LUNAR EXTRAVEHICULAR VISOR

BACKPACK CONTROL BOX

SUNGLASSES POCKET

OXYGEN PURGE SYSTEM ACTUATOR

PORTABLE LIFE SUPPORT SYSTEM

PENLIGHT POCKET

CONNECTOR COVER

COMMUNICATION, VENTILATION, AND LIQUID COOLING UMBILICALS

OXYGEN PURGE SYSTEM UMBILICAL

LM RESTRAINT RING

EXTRAVEHICULAR GLOVE

UTILITY POCKET

INTEGRATED THERMAL METEOROID GARMENT

URINE TRANSFER CONNECTOR, BIOMEDICAL INJECTION, DOSIMETER ACCESS FLAP AND DONNING LANYARD POCKET

LUNAR OVERSHOE

-more-

EXTRAVEHICULAR MOBILITY UNIT

After Apollo 12, a sunshade was added to the outer portion of the LEVA in the middle portion of the helmet rim.

Extravehicular gloves - Built of an outer shell of Chromel-R fabric and thermal insulation the gloves provide protection when handling extremely hot and cold objects. The finger tips are made of silicone rubber to provide more sensitivity.

Constant-wear garment - A one-piece constant-wear garment, similar to "long johns", is worn as an undergarment for the space suit in intravehicular and on CSM EV operations, and with the inflight coveralls. The garment is porous-knit cotton with a waist-to neck zipper for donning. Biomedical harness attach points are provided.

Liquid-cooling garment - The knitted nylon-spandex garment includes a network of plastic tubing through which cooling water from the PLSS is circulated. It is worn next to the skin and replaces the constant-wear garment during lunar surface EVA.

Portable life support system (PLSS) - The backpack supplies oxygen at 3.7 psi and cooling water to the liquid cooling garment. Return oxygen is cleansed of solid and gas contaminants by a lithium hydroxide and activated charcoal canister. The PLSS includes communications and telemetry equipment, displays and controls, and a power supply. The PLSS is covered by a thermal insulation jacket, (two stowed in LM).

Oxygen purge system (OPS) - Mounted atop the PLSS, the oxygen purge system provides a contingency 30-75 minute supply of gaseous oxygen in two bottles pressurized to 5,880 psia, (a minimum of 30 minutes in the maximum flow rate and 75 minutes in the low flow rate). The system may also be worn separately on the front of the pressure garment assembly torso for contingency EVA transfer from the LM to the CSM or behind the neck for CSM EVA. It serves as a mount for the VHF antenna for the PLSS, (two stowed in LM).

Coveralls - During periods out of the space suits, crewmen wear two-piece Teflon fabric inflight coveralls for warmth and for pocket stowage of personal items.

Communications carriers - "Snoopy hats" with redundant microphones and earphones are worn with the pressure helmet; a light-weight headset is worn with the inflight coveralls.

-more-

Water Bags - .9 liter (1 qt.) drinking water bags are attached to the inside neck rings of the EVA suits. The crewman can take a sip of water from the 6-by-8-inch bag through a 1/8-inch-diameter tube within reach of his mouth. The bags are filled from the lunar module potable water dispenser.

Buddy Secondary Life Support System - A connecting hose system which permits a crewman with a failed PLSS to share cooling water in the other crewman's PLSS. The BSLSS lightens the load on the oxygen purge system in the event of a total PLSS failure in that the OPS would supply breathing and pressurizing oxygen while the metabolic heat would be removed by the shared cooling water from the good PLSS. The BSLSS will be stowed on the LRV.

Lunar Boots - The lunar boot is a thermal and abrasion protection device worn over the inner garment and boot assemblies. It is made up of layers of several different materials beginning with Teflon coated beta cloth for the boot liner to Chromel R metal fabric for the outer shell assembly. Aluminized Mylar, Nomex felt, Dacron, Beta cloth and Beta marquisette Kapton comprise the other layers. The lunar boot sole is made of high-strength silicone rubber.

Personal Hygiene

Crew personal hygiene equipment aboard Apollo 16 includes body cleanliness items, the waste management system, and one medical kit.

Packaged with the food are a toothbrush and a two-ounce tube of toothpaste for each crewman. Each man-meal package contains a 3.5-by-4-inch wet-wipe cleansing towel. Additionally, three packages of 12-by-12-inch dry towels are stowed beneath the command module pilot's couch. Each package contains seven towels. Also stowed under the command module pilot's couch are seven tissue dispensers containing 53 three-ply tissues each.

Solid body wastes are collected in plastic defecation bags which contain a germicide to prevent bacteria and gas formation. The bags are sealed after use, identified, and stowed for return to Earth for post-flight analysis.

Urine collection devices are provided for use while wearing either the pressure suit or the inflight coveralls. The urine is dumped overboard through the spacecraft urine dump valve in the CM and stored in the LM. On Apollo 16 urine specimens will be returned to Earth for analysis.

Survival Kit

The survival kit is stowed in two rucksacks in the right-hand forward equipment bay of the CM above the lunar module pilot.

Contents of rucksack No. 1 are: two combination survival lights, one desalter kit, three pairs of sunglasses, one radio beacon, one spare radio beacon battery and spacecraft connector cable, one knife in sheath, three water containers-two containers of Sun lotion, two utility knives, three survival blankets and one utility netting.

Rucksack No. 2: one three-man life raft with CO_2 inflater, one sea anchor, two sea dye markers, three sunbonnets, one mooring lanyard, three manlines and two attach brackets.

The survival kit is designed to provide a 48-hour post-landing (water or land) survival capability for three crewmen between 40 degrees North and South latitudes.

MEDICAL KITS

The command module crew medical supplies are contained in two kits. Included in the larger medical accessories kit are antibiotic ointment, skin cream, eye drops, nose drops, spare biomedical harnesses, oral thermometer and pills of the following types: 18 pain, 12 stimulant, 12 motion sickness, 24 diarrhea, 60 decongestant, 21 sleeping, 72 aspirin and 60 each of two types of antibiotic. A smaller command module auxiliary drug kit contains 80 and 12 of two types of pills for treatment of cardiac arrythymia and two injectors for the same symptom.

The lunar module medical kit contains eye drops, nose drops, antibiotic ointment, bandages and the following pills: 4 stimulant, 4 pain, 8 decongestant, 8 diarrhea, 12 aspirin and 6 sleeping. A smaller kit in the LM contains 8 and 4 units of injectable drugs for cardiac arrythymia and 2 units for pain suppression.

- more -

Crew Food System

The Apollo 16 crew selected menus for their flight from the largest variety of foods ever available for a U.S. manned mission. However, some menu constraints have been imposed by a series of medical requirements. This has limited to some degree the latitude of crew menu selection. For the first time since the flight of Gemini 7 in December 1965, the preflight, inflight, and postflight diets are being precisely controlled to facilitate interpretation of the most extensive series of medical tests to be performed on an Apollo mission.

Menus were designed upon individual crew member physiological requirements in the unique conditions of weightlessness and one-sixth gravity on the lunar surface. Daily menus provide 2750, 2500, and 2650 calories per day for the Commander, Command Module Pilot, and Lunar Module Pilot, respectively.

Food items are assembled into meal units and identified as to crew member and sequence of consumption. Foods stored in a "pantry" section may be used as substitutions for nominal meal items so long as the nutrient intake for a 24-hour period is not altered significantly.

There are various types of food used in the menus. These include freeze-dried rehydratables in spoon-bowl packages; thermostabilized foods (wet packs) in flexible packages and metal easy-open cans; intermediate moisture foods; dry bite-size cubes, and beverages.

Water for drinking and rehydrating food is obtained from two sources in the command module -- a portable dispenser for drinking water and a water spigot at the food preparation station which supplies water at about 145 degrees and 55 degrees Fahrenheit. The potable water dispenser provides a continuous flow of water as long as the trigger is held down, and the food preparation spigot dispenses water in one-ounce increments.

A continuous flow water dispenser similar to the one in the command module is used aboard the lunar module for cold water reconstitution of food stowed aboard the lunar module.

- more -

Water is injected into a food package and the package is kneaded and allowed to sit for several minutes. The bag top is then cut open and the food eaten with a spoon. After a meal, germicide tablets are placed in each bag to prevent fermentation and gas formation. The bags are then rolled and stowed in waste disposal areas in the spacecraft.

Several prototype food packages which are intended for use in Skylab missions in 1973 will be tested by the Apollo 16 crew. These include an improved beverage package design, a salt dispenser, and an improved rehydratable package design. Functional aspects of the package and the behavior of liquid food during extended periods of weightlessness will be observed.

The in-suit drink device, which was used for water previously only on the Apollo 15 mission, will contain a specially formulated beverage powder which will be reconstituted with water prior to extravehicular activities on the lunar surface. This beverage will provide the crewman with fluid, electrolytes (especially potassium), and energy during each of the 3 scheduled EVAs. As on Apollo 15, the crewmen on the lunar surface will have an in-suit food bar to snack on.

New foods for the Apollo 16 mission are limited to thermostabilized ham steak, rehydratable grits, and an intermediate moisture cereal bar.

- more -

APOLLO 16 MENU - BLUE

MEAL	Day 1* 5**, 9, 13**	Day 2, 10	Day 3, 11	Day 4, 8***, 12
A	Peaches+ (WP) Scrambled Eggs (8) (RSB) Bacon Squares (8) (IMB) Grits (RSB) Orange Juice (R) Cocoa (R) (+Peaches-Day 13 (RSB))	Fruit Cocktail (R) Sausage Patties (R) Spiced Fruit Cereal (RSB) Orange Juice (R) Cocoa (R)	Peaches (R) Scrambled Eggs (R) Bacon Squares (8) (RSB) Grits (R) Orange Juice (R) Cocoa (R)	Mixed Fruit+ (WP) Ham Steak (WP) Cornflakes (RSB) White Bread (1), Jelly++ (WP) Orange Juice (R) Cocoa (R) (+Fruit Cocktail-Day 12 ++Delete on Day 12) (R)
B	Chicken & Rice Soup (RSB) Hamburger & White Bread (1) (WP) Pears (IMB) Inst. Breakfast (R) Cereal Bar (DB) Citrus Beverage (R)	Corn Chowder Turkey & Gravey Vanilla Pudding White Bread (1) & Peanut Butter Apple Food Bar (2) Orange Drink	Lobster Bisque (RSB) Bread Rye (2) (WP) Tuna Spread (WP) Cherry Food Bar (2) (IMB) Graham Cracker Cubes (6) (DB) Citrus Beverage (R)	Pea Soup (RSB) Meatballs w/Sauce (WP) Lemon Pudding+ (WP) Sugar Cookies (4) (DB) Peaches (IMB) Orange-Grapefruit Drink (R) (+Pork and Scalloped Potatoes-Day 8 & 12 (RSB))
C	Cr. Tomato Soup (RSB) Spaghetti/Meat Sauce (RSB) Peach Ambrosia (RSB) Apricot Cereal Cubes (4) (DB) Pecans (6) (DB) Cocoa (R)	Cr. Potato Soup Frankfurters (4) Chocolate Pudding Orange-Grapefruit Drink	Romaine Soup (RSB) Beef Steak (WP) Chicken & Rice (RSB) Pin. Fruitcake (4) (DB) Pecans (6) (DB) Grape Drink (R)	Beef & Gravy (WP) Chicken Stew (RSB) Butterscotch Pudding (RSB) Chocolate Bar (DB) Gingerbread (4) (DB) Citrus Beverage (R)

3B SKYLAB MEAL

Turkey & Rice Soup (RC)
Rye Bread (2) (WP)
Chicken Spread (1/3) (WP)
Peaches (CAN)
Peanuts
Orange Beverage (SBD)

DB = Dry Bite
IMB = Intermediate Moisture Bite
R = Rehydratable
RSB = Rehydratable Spoon Bowl
WP = Wet Pack
RC = Rehydratable Can
SBD = Skylab Beverage Dispenser

* Meal C only
** Meal A only
*** Meal B and C only

LM MENU (continued)

APOLLO 16 - LM11 MENU (Blue Velcro LMP, Charles M. Duke)

MEAL	DAY 5		DAY 6		DAY 7		DAY 8	
B	Cream of Tomato Soup	RSB	A Peaches (39 gms)	IMB	A Peaches (39 gms)	IMB	A Peaches (39 gms)	IMB
	Rye Bread (2)		Ham Steak	WP	Beef Steak	WP	Ham Steak	WF
	Tuna Spread	WP	Scrambled Eggs	RSB	Bacon Squares (8)	RSB	Scrambled Eggs	RSE
	Apple Food Bar (2)	IMB	Cinn. Toasted Bread		Spiced Fruit Cereal		Cereal Bar	DE
	Chocolate Bar	DB	Cubes (6)	DB	Instant Breakfast	DB	Apricot Cereal Cubes (6)	DE
	Orange-Grapefruit		Instant Breakfast	R	Orange-Grapefruit	R	Orange Beverage	R
	Beverage	R	Orange-Grapefruit		Beverage		Cocoa	R
			Beverage	R	Cherry Food Bar (2)	IMB		
			Lemon Food Bar (2)	IMB				
C	Shrimp Cocktail	RSB	B Pea Soup	RSB	B Romaine Soup	RSB		
	Turkey & Gravy	WP	Salmon Salad	WP	Tuna Salad	RSB		
	Chocolate Pudding	RSB	Frankfurters (4)	RSB	Meatballs w/Sauce	WP		
	Graham Cracker		Peach Ambrosia		Chicken & Rice	RSB		
	Cubes (6)	DB	Pears	DB	Butterscotch Pudding	IMB		
	Cocoa	R	Cereal Bar	R	Gingerbread (6)	DB		
	Citrus Beverage	R	Citrus Beverage	R	Citrus Beverage	R		
			Cocoa		Cocoa	R		

2-72   APOLLO XVI

PANTRY STOWAGE ITEMS

P/N: 14-0123
Line Items 17W & 17X

BEVERAGES	QTY	SOUPS/SALADS/MEATS	QTY
Cocoa	6	Salmon Salad	3
Coffee (B)	16	Tuna Salad	3
Instant Breakfast	12	Shrimp Cocktail	3
Grapefruit Drink	6	Romaine Soup	3
Orange Beverage	6	Potato Soup	3
Orange-Grapefruit Beverage	6	Pea Soup	3
Orange Juice	12		
Orange-Pineapple Drink	6	Spaghetti w/Meat Sauce	3
		Chicken Stew	3

BREAKFAST ITEMS		SANDWICH SPREADS	
Bacon Squares (8)	6	Peanut Butter	3
Spiced Fruit Cereal	3	Jelly	3
Cornflakes	3	Ham Salad	1
Scrambled Eggs	6		
Grits	3	Catsup*	7
Peach Ambrosia	3	Mustard*	7
Sausage Patties	3		

SNACK ITEMS		ACCESSORIES	
Pecans (6)	3	Wet Skin Cleaning Towels	9
Apricots (IMB)(38.5 gm)	6	Contingency Feeding System	
Peaches (IMB) (39 gm)	8	3 Food Restrainer Pouches	
Pears (IMB) (42 gm)	6	3 Beverage Packages	
Apricot Food Bar (1)(26 gm)	9	1 Valve Adapter (pontube)	
Apple Food Bar (1) (26 gm)	9	Germicidal Tablets	
Lemon Food Bar (1) (26 gm)	9	Index Card	
Cherry Food Bar (1) (26 gm)	9		
Cereal Bar	6		
Chocolate Bar	3		
Sugar Cookies (4)	3		
Graham Crackers (6)	3		
Cheese Cracker Cubes (4)	3		

*Stowage locations TBD

APOLLO 16 FLAGS, LUNAR MODULE PLAQUE

The United States flag to be erected on the lunar surface measures 78 by 125 cm (30 by 48 inches) and will be deployed on a two-piece aluminum tube 2.1 meters (eight feet) long. The nylon flag will be stowed in the lunar module descent stage modularized equipment stowage assembly.

Also carried on the mission and returned to Earth will be 25 United States flags, 50 individual state flags, flags of United States territories, flags of other national states which are generally accepted as independent in the world community, and flags of the United Nations and other international organizations. These flags are 10 by 15 cm (four by six inches).

An 18 by 23 cm (seven by nine inch) stainless steel plaque, similar to that flown on Apollo 15, will be fixed to the LM front leg. The plaque has on it the words "Apollo 16" with "April 1972" and the signatures of the three crewmen located beneath.

SATURN V LAUNCH VEHICLE

The Saturn V launch vehicle (SA-511) assigned to the Apollo 16 mission is very similar to the vehicles used for the missions of Apollo 8 through Apollo 15.

First Stage

The five first stage (S-1C) F-1 engines develop about 34 million newtons (7.7 million pounds) of thrust at launch. Major stage components are the forward skirt, oxidizer tank, intertank structure, fuel tank, and thrust structure. Propellant to the five engines normally flows at a rate of about 13,300 kilograms (29,300 pounds; 3,390 gallons) a second. One engine is rigidly mounted on the stage's centerline; the outer four engines are mounted on a ring at 90-degree angles around the center engine. These outer engines are gimbaled to control the vehicle's attitude during flight.

Second Stage

The five second stage (S-II) J-2 engines develop a total of about 5.15 million newtons (1.15 million pounds) of thrust during flight. Major components are the forward skirt, liquid hydrogen and liquid oxygen tanks (separated by an insulated common bulkhead), a thrust structure, and an interstage section that connects the first and second stages. The engines are mounted and used in the same arrangement as the first stage's F-1 engines: four outer engines can be gimbaled; the center one is fixed.

Third Stage

Major components of the third stage (S-IVB) are a single J-2 engine, aft interstage and skirt, thrust structure, two propellant tanks with a common bulkhead, and forward skirt. The gimbaled engine has a maximum thrust of 1.01 million newtons (230,000 pounds), and can be restarted in Earth orbit.

Instrument Unit

The instrument unit (IU) contains navigation, guidance and control equipment to steer the Saturn V into Earth orbit and translunar trajectory. The six major systems are structural, environmental control, guidance and control, measuring and telemetry, communications, and electrical.

INSTRUMENT UNIT (IU)

Diameter:	6.6 meters (21.7 feet)
Height:	0.9 meters (3 feet)
Weight:	2,040 kilograms (4,500 pounds)

THIRD STAGE (S-IVB)

Diameter:	6.6 meters (21.7 feet)
Height:	18.1 meters (59.3 feet)
Weight:	118,161 kg. fueled (260,500 lbs.)
	11,340 kg. dry (25,100 lbs.)
Engine:	One J-2
Propellants:	Liquid Oxygen (77,590 liters; 20,500 gals.)
	Liquid Hydrogen (243,000 liters; 64,100 gals.)
Thrust:	920,000 newtons (207,000 lbs.)
Interstage:	3,674 kg. (8,100 lbs.)

SECOND STAGE (S-II)

Diameter:	10.1 meters (33 feet)
Height:	24.8 meters (81.5 feet)
Weight:	505,750 kg. fueled (1,115,000 lbs.)
	37,820 kg. dry (83,400 lbs.)
Engines:	Five J-2
Propellants:	Liquid Oxygen (340,270 liters; 89,900 gals.)
	Liquid Hydrogen (1,033,300 liters; 273,000 gals.
Thrust:	5,150,000 newtons (1,150,000 lbs.)
Interstage:	4,581 kg. (10,100 lbs.)

FIRST STAGE (S-IC)

Diameter:	10.1 meters (33 feet)
Height:	42.1 meters (138 feet)
Weight:	2,245,280 kg. fueled (4,950,000 lbs.)
	130,641 kg. dry (288,000 lbs.)
Engines:	Five F-1
Propellants:	Liquid Oxygen (1,300,900 liters; 343,700 gals.)
	RP-1 Kerosene (802,420 liters; 212,000 gals.)
Thrust:	34,356,850 newtons (7,723,726 lbs.) at lift-off

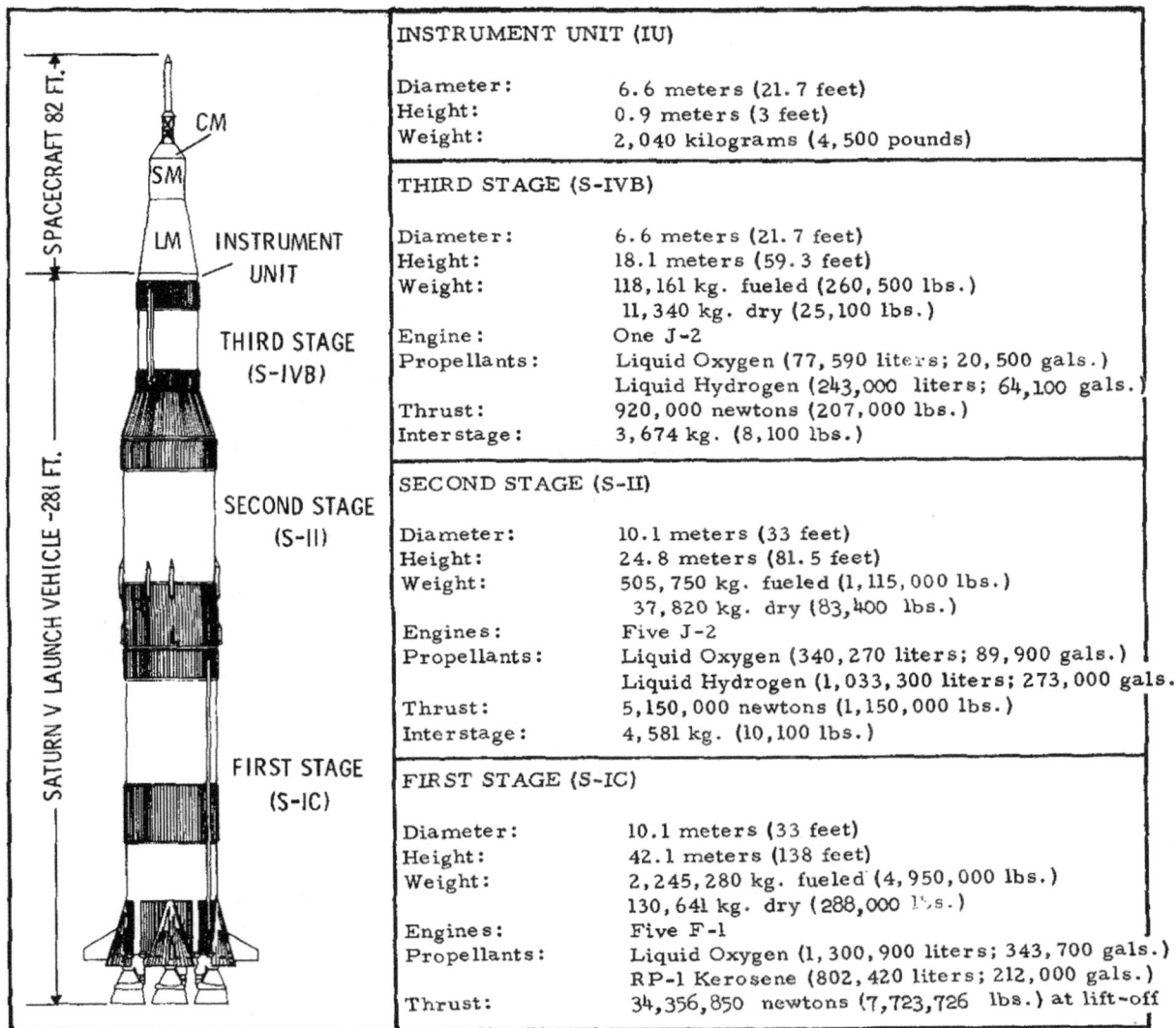

NOTE: Weights and measures given above are for the nominal vehicle configuration for Apollo. The figures may vary slightly due to changes before launch to meet changing conditions. Weights of dry stages and propellants do not equal total weight because frost and miscellaneous smaller items are not included in chart.

SATURN V LAUNCH VEHICLE

-more-

The IU's inertial guidance platform provides space-fixed reference coordinates and measures acceleration during flight. If the platform should fail during boost, systems in the Apollo spacecraft are programmed to provide launch vehicle guidance. After second stage ignition, the spacecraft commander can manually steer the vehicle if its guidance platform is lost.

Propulsion

The Saturn V has 31 propulsive units, with thrust ratings ranging from 311 newtons (70 pounds) to more than 6.8 million newtons (1.53 million pounds). The large main engines burn liquid propellants; the smaller units use solid or hypergolic (self-igniting) propellants.

The five F-1 engines give the first stage a thrust range of from 34,356,850 newtons (7,723,726 pounds) at liftoff to 40,576,160 newtons (9,121,883 pounds) at center engine cutoff. Each F-1 engine weighs almost nine metric tons (10 short tons), is more than 5.5 meters long (18 feet), and has a nozzle exit diameter of nearly 4.6 meters (14 feet). Each engine uses almost 2.7 metric tons (3 short tons) of propellant a second.

J-2 engine thrust on the second and third stages averages 960,000 newtons (216,000 pounds) and 911,000 newtons (205,000 pounds) respectively during flight. The 1,590-kilogram (3,500-pound) engine uses high-energy, low-molecular-weight liquid hydrogen as propellant.

The first stage has eight solid-fuel retro-rockets that fire to separate the first and second stages. Each rocket produces a thrust of 237,000 newtons (75,800 pounds) for 0.54 seconds.

Four retrorockets, located in the third stage's aft interstage, separate the second and third stages. Two jettisonable ullage rockets settle propellants before engine ignition. Six smaller engines in two auxiliary propulsion system modules on the third stage provide three-axis attitude control.

Significant Vehicle Changes

Saturn vehicle SA-511 is similar in configuration to the Apollo 15 launch vehicle. The first stage (S-1C) has eight retrorocket motors, double the number on the SA-510 vehicle, because flight evaluation of the Apollo 15 mission revealed that the separation distance between the first and second stages was less than predicted. Eight retrorockets will give a greater safety margin should one motor fail during separation.

APOLLO SPACECRAFT

The Apollo spacecraft consists of the command module, service module, lunar module, a spacecraft lunar module adapter (SLA), and a launch escape system. The SLA houses the lunar module and serves as a mating structure between the Saturn V instrument unit and the SM.

Launch Escape System (LES) -- The function of the LES is to propel the command module to safety in an aborted launch. It has three solid-propellant rocket motors: a 658,000-newton (147,000-pound)-thrust launch escape system motor, a 10,750-newton (2,400-pound)-thrust pitch control motor, and a 141,000 newton (31,500-pound)-thrust tower jettison motor. Two canard vanes deploy to turn the command module aerodynamically to an attitude with the heat-shield forward. The system is 10 meters (33 feet) tall and 1.2 meters (four feet) in diameter at the base, and weighs 4,158 kilograms (9,167 pounds).

Command Module (CM) -- The command module is a pressure vessel encased in heat shields, cone-shaped, weighing 5,843.9 kg (12,874 lb.) at launch.

The command module consists of a forward compartment which contains two reaction control engines and components of the Earth landing system; the crew compartment or inner pressure vessel containing crew accommodations, controls and displays, and many of the spacecraft systems; and the aft compartment housing ten reaction control engines, propellant tankage, helium tanks, water tanks, and the CSM umbilical cable. The crew compartment contains 6 m^3 (210 ft.^3) of habitable volume.

Heat-shields around the three compartments are made of brazed stainless steel honeycomb with an outer layer of phenolic epoxy resin as an ablative material.

The CSM and LM are equipped with the probe-and-drogue docking hardware. The probe assembly is a powered folding coupling and impact attentuating device mounted in the CM tunnel that mates with a conical drogue mounted in the LM docking tunnel. After the 12 automatic docking latches are checked following a docking maneuver, both the probe and drogue are removed to allow crew transfer between the CSM and LM.

- more -

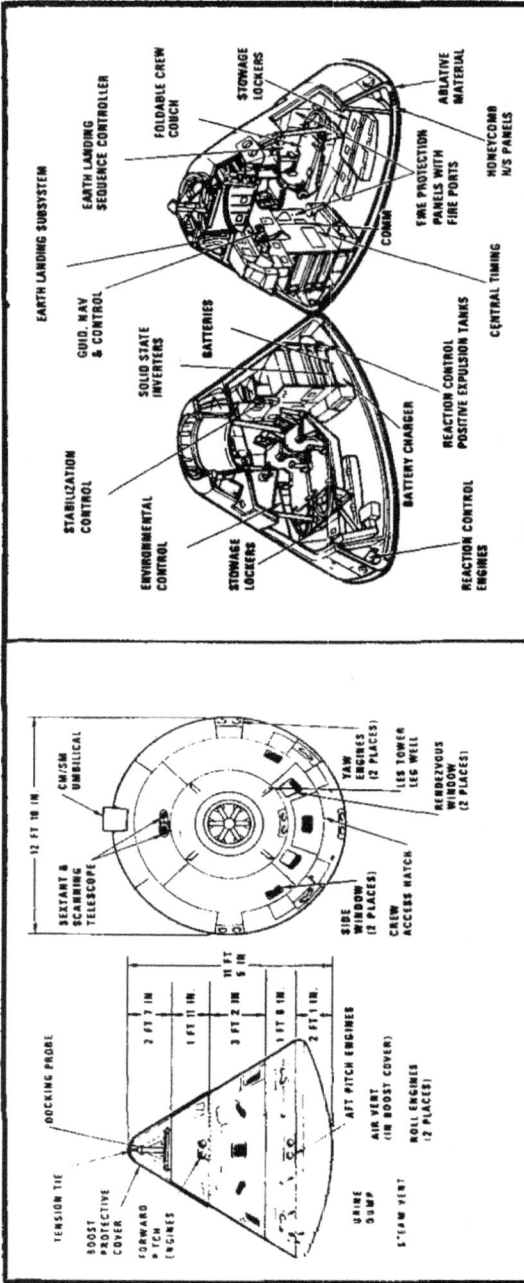

COMMAND MODULE

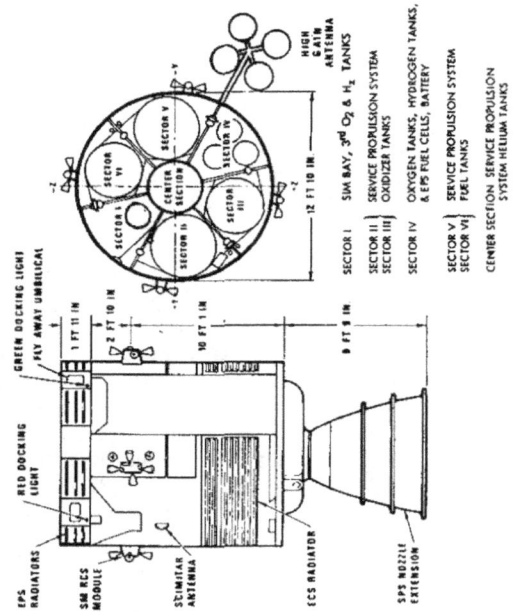

EARTH LANDING SUBSYSTEM

EARTH LANDING SEQUENCE CONTROLLER

FOLDABLE CREW COUCH

STOWAGE LOCKERS

ABLATIVE MATERIAL

HONEYCOMB N/S PANELS

COMM

FIRE PROTECTION PANELS WITH FIRE PORTS

CENTRAL TIMING

REACTION CONTROL POSITIVE EXPULSION TANKS

GUID, NAV & CONTROL

SOLID STATE INVERTERS

BATTERIES

BATTERY CHARGER

STABILIZATION CONTROL

ENVIRONMENTAL CONTROL

STOWAGE LOCKERS

REACTION CONTROL ENGINES

12 FT 10 IN

SEXTANT & SCANNING TELESCOPE

CM/SM UMBILICAL

YAW ENGINES (2 PLACES)

LES TOWER LEG WELL

RENDEZVOUS WINDOW (2 PLACES)

SIDE WINDOW (2 PLACES)

CREW ACCESS HATCH

11 FT 5 IN

2 FT 1 IN

1 FT 11 IN

3 FT 2 IN

1 FT 8 IN

2 FT 1 IN.

DOCKING PROBE

TENSION TIE

BOOST PROTECTIVE COVER

FORWARD PITCH ENGINES

URINE DUMP

STEAM VENT

AFT PITCH ENGINES

AIR VENT (IN BOOST COVER)

ROLL ENGINES (2 PLACES)

SERVICE MODULE

HIGH GAIN ANTENNA

SECTOR V

SECTOR VI

SECTOR I

SECTOR IV

CENTER SECTION

SECTOR III

SECTOR II

12 FT 10 IN

SECTOR I   SIM BAY, 3rd O_2 & H_2 TANKS

SECTOR II   SERVICE PROPULSION SYSTEM
SECTOR III   OXIDIZER TANKS

SECTOR IV   OXYGEN TANKS, HYDROGEN TANKS,
                & EPS FUEL CELLS, BATTERY

SECTOR V }   SERVICE PROPULSION SYSTEM
SECTOR VI}   FUEL TANKS

CENTER SECTION SERVICE PROPULSION
                SYSTEM HELIUM TANKS

GREEN DOCKING LIGHT

FLY AWAY UMBILICAL

1 FT 11 IN

2 FT 10 IN

10 FT 1 IN

9 FT 9 IN

RED DOCKING LIGHT

EPS RADIATORS

SM RCS MODULE

SCIMITAR ANTENNA

ECS RADIATOR

SPS NOZZLE EXTENSION

Service Module (SM) -- The Apollo 16 service module will weigh 24,514 kg (54,044 lb.) at launch, of which 18,415 kg (40,594 lb.) is propellant for the 91,840-newton (20,500-pound)-thrust service propulsion engine: (fuel: 50/50 hydrazine and unsymmetrical dimethyl-hydrazine; oxidizer: nitrogen tetroxide). Aluminum honeycomb panels 2.54 centimeters (one inch) thick form the outer skin, and milled aluminum radial beams separate the interior into six sections around a central cylinder containing service propulsion system (SPS) helium pressurant tanks. The six sectors of the service module house the following components: Sector I -- oxygen tank 3 and hydrogen tank 3, J-mission SIM bay; Sector II -- space radiator, +Y RCS package, SPS oxidizer storage tank; Sector III -- space radiator, +Z RCS package, SPS oxidizer storage tank; Sector IV -- three fuel cells, two oxygen tanks, two hydrogen tanks, auxiliary battery; Sector V -- space radiator, SPS fuel sump tank, -Y RCS package; Sector VI -- space radiator, SPS fuel storage tank, -Z RCS package.

Spacecraft-LM adapter (SLA)Structure -- The spacecraft-LM adapter is a truncated cone 8.5 m (28 ft.) long tapering from 6.7 m (21.6 ft.) in diameter at the base to 3.9 m (12.8 ft.) at the forward end at the service module mating line. The SLA weighs 1,841 kg (4,059 lb.) and houses the LM during launch and the translunar injection maneuver until CSM separation, transposition, and LM extraction. The SLA quarter panels are jettisoned at CSM separation.

Lunar Module (LM)

The lunar module is a two-stage vehicle designed for space operations near and on the Moon. The lunar module stands 7 m (22 ft. 11 in.) high and is 9.5 m (31 ft.) wide (diagonally across landing gear). The ascent and descent stages of the LM operate as a unit until staging, when the ascent stage functions as a single spacecraft for rendezvous and docking with the CM.

Ascent Stage -- Three main sections make up the ascent stage: the crew compartment, midsection, and aft equipment bay. Only the crew compartment and midsection are pressurized 337.5 grams per square centimeter (4.8 pounds per square inch gauge). The cabin volume is 6.7 cubic meters (235 cubic feet). The stage measures 3.8 m (12 ft. 4 in.) high by 4.3 m (14 ft. 1 in.) in diameter. The ascent stage has six substructural areas: crew compartment, midsection, aft equipment bay, thrust chamber assembly cluster supports, antenna supports, and thermal and micro-meteoroid shield.

The cylindrical crew compartment is 2.35 m (7 ft. 10 in.) in diameter and 1.07 m (3 ft. 6 in.) deep. Two flight stations are equipped with control and display panels, armrests, body restraints, landing aids, two front windows, an overhead docking window, and an alignment optical tele-scope in the center between the two flight stations. The habitable volume is 4.5 m^3 (160 ft.^{3}).

A tunnel ring atop the ascent stage meshes with the command module docking latch assemblies. During docking, the CM docking ring and latches are aligned by the LM drogue and the CSM probe.

The docking tunnel extends downward into the mid-section 40 cm (16 in.). The tunnel is 81 cm (32 in.) in diameter and is used for crew transfer between the CSM and LM. The upper hatch on the inboard end of the docking tunnel opens inward and cannot be opened without equalizing pressure on both hatch surfaces.

A thermal and micrometeoroid shield of multiple layers of Mylar and a single thickness of thin aluminum skin encases the entire ascent stage structure.

- more -

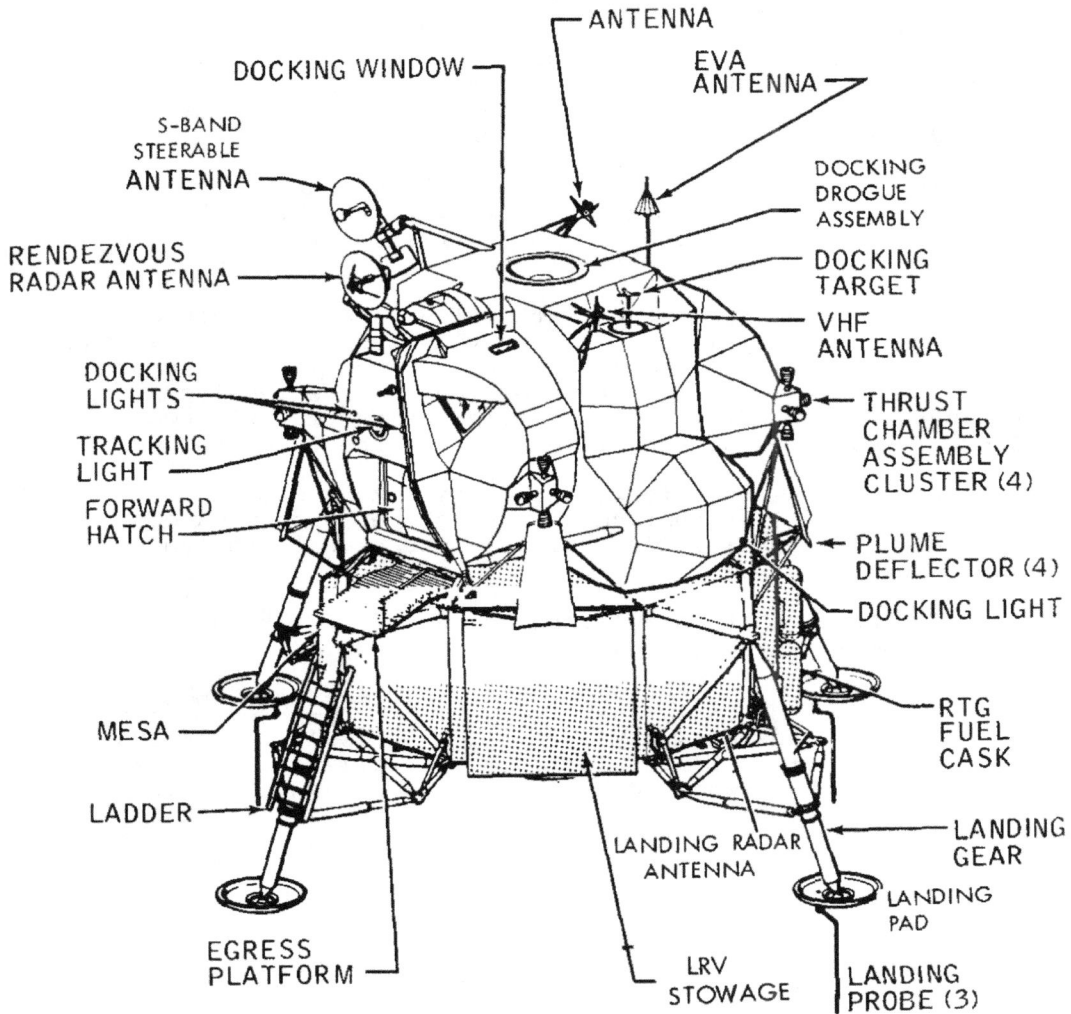

ANTENNA

EVA ANTENNA

DOCKING WINDOW

S-BAND STEERABLE ANTENNA

RENDEZVOUS RADAR ANTENNA

DOCKING DROGUE ASSEMBLY

DOCKING TARGET

VHF ANTENNA

DOCKING LIGHTS

TRACKING LIGHT

FORWARD HATCH

THRUST CHAMBER ASSEMBLY CLUSTER (4)

PLUME DEFLECTOR (4)

DOCKING LIGHT

MESA

LADDER

RTG FUEL CASK

LANDING GEAR

LANDING PAD

EGRESS PLATFORM

LANDING RADAR ANTENNA

LRV STOWAGE

LANDING PROBE (3)

LUNAR MODULE

-more-

Descent Stage -- The descent stage center compartment
houses the descent engine, and descent propellant tanks
are housed in the four bays around the engine. Quadrant
II contains ALSEP. The radioisotope thermoelectric generator (RTG) is externally mounted. Quadrant IV contains the
MESA. The descent stage measures 3.2 m (10 ft. 7 in.)
high by 4.3 m (14 ft. 1 in.) in diameter and is encased
in the Mylar and aluminum alloy thermal and micrometeoroid
shield. The LRV is stowed in Quadrant I.

The LM egress platform or "porch" is mounted on the
forward outrigger just below the forward hatch. A ladder
extends down the forward landing gear strut from the porch
for crew lunar surface operations.

The landing gear struts are released explosively and
are extended by springs. They provide lunar surface landing
impact attenuation. The main struts are filled with crush-
able aluminum honeycomb for absorbing compression loads.
Footpads 0.95 m (37 in.) in diameter at the end of each
landing gear provide vehicle support on the lunar surface.

Each pad (except forward pad) is fitted with a 1.7-m
(68-in.) long lunar surface sensing probe which upon contact with the lunar surface signals the crew to shut down
the descent engine.

The Apollo LM has a launch weight of 16,429 kg (36,218
lb.). The weight breakdown is as follows:

		kilograms	pounds
1.	Ascent stage, dry*	2,134	4,704
2.	APS propellants (loaded)	2,378	5,242
3.	Descent stage, dry	2,759	6,083
4.	DPS propellants (loaded)	8,872	19,558
5.	RCS propellants (loaded)	286	631
		16,429 kg	36,218 lbs

* Includes water and oxygen; no crew.

- more -

NATIONAL AERONAUTICS AND SPACE ADMINISTRATION

WASHINGTON, D. C. 20546

BIOGRAPHICAL DATA

NAME: John W. Young (Captain, USN)
Apollo 16 Mission Commander
NASA Astronaut

BIRTHPLACE AND DATE: Born in San Francisco, California,
on September 24, 1930. His parents, Mr. and Mrs.
William H. Young, reside in Orlando, Florida.

PHYSICAL DESCRIPTION: Brown hair; green eyes, height:
5 feet 9 inches; weight: 165 pounds.

EDUCATION: Graduated from Orlando High School. Orlando,
Florida; received a Bachelor of Science degree in
Aeronautical Engineering from the Georgia Institute
of Technology in 1952; recipient of an Honorary
Doctorate of Laws degree from Western State Univer-
sity College of Law in 1969, and an Honorary Doc-
torate of Applied Science from Florida Technological
University in 1970.

MARITAL STATUS: Married to the former Susy Feldman of
St. Louis, Missouri.

CHILDREN: Sandy, April 30, 1957; John, January 17, 1959
(by a previous marriage).

RECREATIONAL INTERESTS: He plays handball, runs and works
out in the full pressure suit to stay in shape.

ORGANIZATIONS: Fellow of the American Astronautical
Society, Associate Fellow of the Society of
Experimental Test Pilots, and Member of the
American Institute of Aeronautics and Astronautics.

SPECIAL HONORS: Awarded the NASA Distinguished Service
Medal, two NASA Exceptional Service Medals, the
MSC Certificate of Commendation (1970), the Navy
Astronaut Wings, the Navy Distinguished Service
Medals, and three Navy Distinguished Flying Crosses.

- more -

EXPERIENCE: Upon graduation from Georgia Tech, Young entered the U.S. Navy in 1952; he holds the rank of Captain in that service.

He completed test pilot training at the U.S. Naval Test Pilot School in 1959, and was then assigned as a test pilot at the Naval Air Test Center until 1962. Test projects in which he participated include evaluations of the F8D "Crusader" and F4B "Phantom" fighter weapons systems, and in 1962, he set world time-to-climb records to 3,000 and 25,000 meter altitudes in the Phantom. Prior to his assignment to NASA, he was maintenance officer of All-Weather-Fighter Squadron 143 at the Naval Air Station, Miramar, California. He has logged more than 5,900 hours flying time, including more than 4,900 hours in jet aircraft, and completed three space flights totaling 267 hours and 42 minutes. Captain Young was selected as an astronaut by NASA in September 1962. He served as pilot with command pilot Gus Grissom on the first manned Gemini flight -- a 3-orbit mission, launched on March 23, 1965, during which the crew accomplished the first manned space-craft orbital trajectory modifications and lifting reentry, and flight tested all systems in Gemini 3.

After this flight, he was backup pilot for Gemini 6.

On July 18, 1966, Young occupied the command pilot seat for the Gemini 10 mission and, with Michael Collins as pilot, effected a successful rendezvous and docking with the Agena target vehicle. The large Agena main engine was subsequently ignited, propelling the docked combination to a record altitude of approximately 475 miles above the earth, the first manned operation of a large rocket engine in space. They later performed a completely optical rendezvous (without radar) on a second, passive Agena which had been placed in orbit during the Gemini 8 mission. After the rendezvous, Young flew formation on the slowly rotating passive Agena while Collins performed extra-vehicular activity to it and recovered a micrometeorite detection experiment, accomplishing an in-space retrieval of the detector which had been orbiting the earth for three months. The flight was concluded, after 3 days and 44 revolutions, with a precise splashdown in the West Atlantic, 2.6 miles from the recovery ship USS GUADALCANAL.

- more -

He was then assigned as the backup command module pilot for Apollo 7.

Young was command module pilot for Apollo 10, May 18-26, 1969, the comprehensive lunar-orbital qualification test of the Apollo lunar module. He was accompanied on the 248,000 nautical mile lunar mission by Thomas P. Stafford (spacecraft commander) and Eugene A. Cernan (lunar module pilot). In achieving all mission objectives, Apollo 10 confirmed the operational performance, stability, and reliability of the command-service module/lunar module configuration during translunar coast, lunar orbit insertion, and lunar module separation and descent to within 8 nautical miles of the lunar surface. The latter maneuvers employed all but the final landing phase and facilitated extensive evaluations of the lunar module landing radar devices and propulsion systems and lunar module and command-service module rendezvous radars in the subsequent lunar rendezvous. In addition to discovering unexpectedly that the lunar gravitational field targeted the lunar module over 4 miles south of the Apollo 11 lunar landing site, Apollo 10 photographed and accurately located this site for the lunar landing.

Captain Young served as backup spacecraft commander for Apollo 13.

He was assigned as spacecraft commander for the Apollo 16 flight March 3, 1971.

- end -

March 1972

NATIONAL AERONAUTICS AND SPACE ADMINISTRATION

WASHINGTON, D. C. 20546

BIOGRAPHICAL DATA

NAME: Thomas K. Mattingly II (Lieutenant Commander, USN)
        Apollo 16 Command Module Pilot
        NASA Astronaut

BIRTHPLACE AND DATE: Born in Chicago, Illinois, March 17,
        1936. His parents, Mr. and Mrs. Thomas K. Mattingly,
        now reside in Hialeah, Florida.

PHYSICAL DESCRIPTION: Brown hair; blue eyes; height:
        5 feet 10 inches; weight: 140 pounds.

EDUCATION: Attended Florida elementary and secondary
        schools and is a graduate of Miami Edison High
        School, Miami, Florida; received a Bachelor of
        Science degree in Aeronautical Engineering from
        Auburn University in 1958.

MARITAL STATUS: Married to the former Elizabeth Dailey
        of Hollywood, California.

RECREATIONAL INTERESTS: Enjoys water skiing and playing
        handball and tennis.

ORGANIZATIONS: Member of the American Institute of
        Aeronautics and Astronautics and the U.S. Naval
        Institute.

SPECIAL HONORS: Presented the MSC Certificate of Commen-
        dation (1970).

EXPERIENCE: Prior to reporting for duty at the Manned
        Spacecraft Center, he was a student at the Air
        Force Aerospace Research Pilot School.

        He began his Naval career as an Ensign in 1958 and
        received his wings in 1960. He was then assigned
        to VA-35 and flew A1H aircraft aboard the USS
        SARATOGA from 1960 to 1963. In July 1963, he
        served in VAH-11 deployed aboard the USS FRANKLIN D.
        ROOSEVELT where he flew the A3B aircraft for two years.

        He logged 4,200 hours of flight time -- 2,300 hours
        in jet aircraft.

- more -

Lt. Commander Mattingly is one of the 19 astronauts selected by NASA in April 1966. He served as
a member of the astronaut support crews for the Apollo
8 and 9 missions.

He was then designated command module pilot for
Apollo 13 but was removed from flight status 72
hours prior to the scheduled launch due to exposure
to the german measles. Mattingly was replaced on
the flight of Apollo 13, April 13-17, 1970, by
backup command module pilot John L. Swigert, Jr.

He was designated to serve as command module pilot
for the Apollo 16 flight March 3, 1971.

- end -

March 1972

NATIONAL AERONAUTICS AND SPACE ADMINISTRATION

WASHINGTON, D. C. 20546

BIOGRAPHICAL DATA

NAME: Charles Moss Duke, Jr. (Lieutenant Colonel, USAF)
Apollo 15 Lunar Module Pilot
NASA Astronaut

BIRTHPLACE AND DATE: Born in Charlotte, North Carolina,
on October 3, 1935. His parents, Mr. and Mrs.
Charles M. Duke, make their home in Lancaster,
South Carolina.

PHYSICAL DESCRIPTION: Brown hair; brown eyes; height:
5 feet 11 1/2 inches; weight: 155 pounds.

EDUCATION: Attended Lancaster High School in Lancaster,
South Carolina, and was graduated valedictorian
from the Admiral Farragut Academy in St. Petersburg,
Florida; received a Bachelor of Science degree in
Naval Sciences from the U.S. Naval Academy in 1957
and a Master of Science degree in Aeronautics and
Astronautics from the Massachusetts Institute of
Technology in 1964.

MARITAL STATUS: Married to the former Dorothy Meade
Claiborne of Atlanta, Georgia; her parents are
Dr. and Mrs. T. Sterling Claiborne of Atlanta.

CHILDREN: Charles M., March 8, 1965; Thomas C., May 1, 1967.

RECREATIONAL INTERESTS: Hobbies include hunting,
fishing, reading, and playing golf.

ORGANIZATIONS: Member of the Air Force Association, the
Society of Experimental Test Pilots, the Rotary
Club, the American Legion, and the American Fighter
Pilots Association.

SPECIAL HONORS: Awarded the MSC Certificate of Commendation (1970).

- more -

EXPERIENCE: When notified of his selection as an astronaut, Duke was at the Air Force Aerospace Research Pilot School as an instructor teaching control systems and flying in the F-104, F-101, and T-33 aircraft. He was graduated from the Aerospace Research Pilot School in September 1965 and stayed on there as an instructor.

He is an Air Force Lt. Colonel and was commissioned in 1957 upon graduation from the Naval Academy. Upon entering the Air Force, he went to Spence Air Base, Georgia, for primary flight training and then to Webb Air Force Base, Texas, for basic flying training, where in 1958 he became a distinguished graduate. He was again a distinguished graduate at Moody Air Force Base, Georgia, where he completed advanced training in F-86L aircraft. Upon completion of this training he was assigned to the 526th Fighter Interceptor Squadron at Ramstein Air Base, Germany, where he served three years as a fighter interceptor pilot.

He has logged 3,000 hours flying time, which includes 2,750 hours in jet aircraft.

Lt. Colonel Duke is one of the 19 astronauts selected by NASA in April 1966. He served as a member of the astronaut support crew for the Apollo 10 flight and as backup lunar module pilot for the Apollo 13 flight.

He was designated March 3, 1971 to serve as lunar module pilot for the Apollo 16 mission.

- end -

NATIONAL AERONAUTICS AND SPACE ADMINISTRATION

WASHINGTON, D. C. 20546

BIOGRAPHICAL DATA

NAME: Fred Wallace Haise, Jr. (Mr.)
Apollo 16 Backup Commander
NASA Astronaut

BIRTHPLACE AND DATE: Born in Biloxi, Mississippi, on
November 14, 1933; his mother, Mrs. Fred W. Haise, Sr.,
resides in Biloxi.

PHYSICAL DESCRIPTION: Brown hair; brown eyes, height:
5 feet 9 1/2 inches; weight: 155 pounds.

EDUCATION: Graduated from Biloxi High School, Biloxi,
Mississippi; attended Perkinston Junior College
(Association of Arts); received a Bachelor of
Science degree with honors in Aeronautical Engineering
from the University of Oklahoma in 1959, and an
Honorary Doctorate of Science from Western Michigan
University in 1970.

MARITAL STATUS: Married to the former Mary Griffin Grant
of Biloxi, Mississippi. Her parents, Mr. and Mrs.
William J. Grant, Jr., reside in Biloxi.

CHILDREN: Mary M., January 25, 1956; Frederick T.,
May 13, 1958; Stephen W., June 30, 1961; Thomas J.,
July 6, 1970.

ORGANIZATIONS: Fellow of the American Astronautical Society,
and member of the Society of Experimental Test
Pilots, Tau Beta Pi, Sigma Gamma Tau, and Phi
Theta Kappa.

SPECIAL HONORS: Awarded the Presidential Medal for Freedom
(1970), the NASA Distinguished Service Medal, the
AIAA Haley Astronautics Award for 1971, the American
Astronautical Society Flight Achievement Award for
1970, the City of New York Gold Medal in 1970, the
City of Houston Medal for Valor in 1970, the Jeff
Davis Award (1970), the Mississippi Distinguished
Civilian Service Medal (1970), the American Defense
Ribbon, the Society of Experimental Test Pilots Ray E.
Tenhoff Award for 1966, and the A.B. Honts Trophy as
the outstanding graduate of Class 64A from the Aerospace
Research Pilot School in 1964.

- more -

EXPERIENCE: Haise was a research pilot at the NASA Flight
       Research Center at Edwards, California, before
       coming to Houston and the Manned Spacecraft Center;
       and from September 1959 to March 1963, he was a
       research pilot at the NASA Lewis Research Center in
       Cleveland, Ohio. During this time he authored the
       following papers which have been published: a
       NASA TND, entitled "An Evaluation of the Flying
       Qualities of Seven General-Aviation Aircraft";
       NASA TND 3380, "Use of Aircraft for Zero Gravity
       Environment, May 1966"; SAE Business Aircraft Con-
       ference Paper, entitled "An Evaluation of General-
       Aviation Aircraft Flying Qualities, March 30 -
       April 1, 1966"; and a paper delivered at the Tenth
       Symposium of the Society of Experimental Test
       Pilots, entitled "A Quantitative/Qualitative
       Handling Qualities Evaluation of Seven General-
       Aviation Aircraft, 1966."

He was the Aerospace Research Pilot School's
outstanding graduate of Class 64A and served with
the U.S. Air Force from October 1961 to August 1962
as a tactical fighter pilot and as chief of the
164th Standardization-Evaluation Flight of the
164th Tactical Fighter Squadron at Mansfield, Ohio.
From March 1957 to September 1959, Haise was a
fighter interceptor pilot with the 185th Fighter
Interceptor Squadron in the Oklahoma Air National
Guard.

He also served as a tactics and all weather flight
instructor in the U.S. Navy Advanced Training
Command at NAAS Kingsville, Texas, and was assigned
as a U.S. Marine Corps fighter pilot to VMF-533
and 114 at MCAS Cherry Point, North Carolina, from
March 1954 to September 1956.

His military career began in October 1952 as a
Naval Aviation Cadet at the Naval Air Station in
Pensacola, Florida.

He has accumulated 6,700 hours flying time, including
3,300 hours in jets.

- more -

Mr. Haise is one of the 19 astronauts selected by
NASA in April 1966. He served as backup lunar
module pilot for the Apollo 8 and 11 missions.

Haise was lunar module pilot for Apollo 13, April 11-17.
1970. Apollo 13 was programmed for ten days and
was committed to our first landing in the hilly,
upland Fra Mauro region of the Moon; however, the
original flight plan was modified enroute to the
Moon due to a failure of the service module
cryogenic oxygen system which occurred at approximately
55 hours into the flight. Haise and fellow crewmen,
James A. Lovell (spacecraft commander) and John L.
Swigert (command module pilot), working closely
with Houston ground controllers, converted their
lunar module "Aquarius" into an effective lifeboat.
Their emergency activation and operation of lunar
module systems conserved both electrical power and
water in sufficient supply to assure their safety
and survival while in space and for the return to
Earth.

In completing his first space flight, Mr. Haise
logged a total of 142 hours and 54 minutes in
space.

He was designated as backup spacecraft commander
for the Apollo 16 mission March 3, 1971.

- end -

March 1972

NATIONAL AERONAUTICS AND SPACE ADMINISTRATION

WASHINGTON, D. C. 20546

BIOGRAPHICAL DATA

NAME: Stuart Allen Roosa (Lieutenant Colonel, USAF)
Apollo 16 Backup Command Module Pilot
NASA Astronaut

BIRTHPLACE AND DATE: Born August 16, 1933, in Durango,
Colorado. His parents, Mr. and Mrs. Dewey Roosa,
now reside in Tucson, Arizona.

PHYSICAL DESCRIPTION: Red hair; blue eyes, height:
5 feet 10 inches; weight: 155 pounds.

EDUCATION: Attended Justice Grade School and Claremore
High School in Claremore, Oklahoma; studied at
Oklahoma State University and the University of
Arizona and was graduated with honors and a
Bachelor of Science degree in Aeronautical Engineering
from the University of Colorado; presented an
Honorary Doctorate of Letters from the University
of St. Thomas (Houston, Texas) in 1971.

MARITAL STATUS: His wife is the former Joan C. Barrett
of Tupello, Mississippi; and her mother, Mrs. John T.
Barrett, resides in Sessums, Mississippi.

CHILDREN: Christopher A., June 29, 1959; John D.,
January 2, 1961; Stuart A., Jr., March 12, 1962;
Rosemary D., July 23, 1963.

RECREATIONAL INTERESTS: His hobbies are hunting, boating,
and fishing.

ORGANIZATIONS: Associate Member of the Society of
Experimental Test Pilots.

SPECIAL HONORS: Presented the NASA Distinguished Service
Medal, the MSC Superior Achievement Award (1970),
the Air Force Command Pilot Astronaut Wings, the
Air Force Distinguished Service Medal, the Arnold
Air Society's John F. Kennedy Award (1971), and
the City of New York Gold Medal in 1971.

- more -

EXPERIENCE: Roosa, a lt. colonel in the Air Force, has been
on active duty since 1953. Prior to joining NASA, he
was an experimental test pilot at Edwards Air Force
Base, Calif. -- an assignment he held from September
1965 to May 1966, following graduation from the Aerospace Research Pilots School. He was a maintenance
flight test pilot at Olmstead Air Force Base, Pennsylvania, from July 1962 to August 1964, flying F-101
aircraft. He served as Chief of Service Engineering
(AFLC) at Tachikawa Air Base for two years following
graduation from the University of Colorado under the
Air Force Institute of Technology Program. Prior to
this tour of duty, he was assigned as a fighter pilot
at Langley Air Force Base, Virginia, where he flew
the F-84F and F-100 aircraft.

He attended Gunnery School at Del Rio and Luke Air
Force Bases and is a graduate of the Aviation Cadet
Program at Williams Air Force Base, Arizona, where
he received his flight training and commission in the
Air Force. Since 1953, he has acquired 4,600 flying
hours -- 4,100 hours in jet aircraft.

Lt. Colonel Roosa is one of the 19 astronauts selected
by NASA in April 1966. He was a member of the astronaut support crew for the Apollo 9 flight.

He completed his first space flight as command module
pilot on Apollo 14, January 31-February 9, 1971. With
him on man's third lunar landing mission were Alan B.
Shepard (spacecraft commander) and Edgar D. Mitchell
(lunar module pilot).

Maneuvering their lunar module, "Antares," to a landing
in the hilly upland Fra Mauro region of the moon, Shepar
and Mitchell subsequently deployed and activated various
scientific equipment and experiments and proceeded to
collect almost 100 pounds of lunar samples for return
to earth. Throughout this 33-hour period of lunar surface activities, Roosa remained in lunar orbit aboard
the command module, "Kittyhawk," to conduct a variety
of assigned photographic and visual observations.
Apollo 14 achievements include: first use of Mobile
Equipment Transporter (MET); largest payload placed in
lunar orbit; longest distance traversed on the lunar
surface; largest payload returned from the lunar surface; longest lunar surface stay time (33 hours);
longest lunar surface EVA (9 hours and 17 minutes);
first use of shortened lunar orbit rendezvous techniques;
first use of colored TV with new vidicon tube on
lunar surface; and first extensive orbital science
period conducted during CSM solo operations.

- more -

In completing his first space flight, Roosa logged a total of 216 hours and 42 minutes.

He was designated to serve as backup command module pilot for Apollo 16 on March 3, 1971.

- end -

March 1972

NATIONAL AERONAUTICS AND SPACE ADMINISTRATION

WASHINGTON, D. C. 20546

BIOGRAPHICAL DATA

NAME: Edgar Dean Mitchell (Captain, USN)
Apollo 16 Backup Lunar Module Pilot
NASA Astronaut

BIRTHPLACE AND DATE: Born in Hereford, Texas, on September 17, 1930, but considers Artesia, New Mexico, his hometown. His mother, Mrs. J. T. Mitchell, now resides in Tahlequah, Oklahoma.

PHYSICAL DESCRIPTION: Brown hair; green eyes; height: 5 feet 11 inches; weight: 180 pounds.

EDUCATION: Attended primary schools in Roswell, New Mexico, and is a graduate of Artesia High School in Artesia, New Mexico; received a Bachelor of Science degree in Industrial Management from the Carnegie Institute of Technology in 1952, a Bachelor of Science degree in Aeronautical Engineering from the U.S. Naval Postgraduate School in 1961, and a Doctorate of Science degree in Aeronautics/Astronautics from the Massachusetts Institute of Technology in 1964; presented an Honorary Doctorate of Science from New Mexico State University in 1971, and an Honorary Doctorate of Engineering from Carnegie-Mellon University in 1971.

MARITAL STATUS: Married to the former Louise Elizabeth Randall of Muskegon, Michigan. Her mother, Mrs. Winslow Randall, now resides in Pittsburgh, Pennsylvania.

CHILDREN: Karlyn L., August 12, 1953; Elizabeth R., March 24, 1959.

RECREATIONAL INTERESTS: He enjoys handball and swimming, and his hobbies are scuba diving and soaring.

ORGANIZATIONS: Member of the American Institute of Aeronautics and Astronautics; the Society of Experimental Test Pilots; Sigma Xi; and Sigma Gamma Tau.

- more -

SPECIAL HONORS: Presented the Presidential Medal of Freedom
        (1970), the NASA Distinguished Service Medal, the
        MSC Superior Achievement Award (1970), the Navy
        Astronaut Wings, the Navy Distinguished Service
        Medal, the City of New York Gold Medal (1971), and
        the Arnold Air Society's John F. Kennedy Award (1971).

EXPERIENCE: Captain Mitchell's experience includes Navy
        operational flight, test flight, engineering,
        engineering management, and experience as a college
        instructor. Mitchell came to the Manned Spacecraft
        Center after graduating first in his class from
        the Air Force Aerospace Research Pilot School where
        he was both student and instructor.

        He entered the Navy in 1952 and completed his basic
        training at the San Diego Recruit Depot. In May 1953,
        after completing instruction at the Officers' Candidate
        School at Newport, Rhode Island, he was commissioned
        as an ensign. He completed flight training in
        July 1954 at Hutchinson, Kansas, and subsequently
        was assigned to Patrol Squadron 29 deployed to
        Okinawa.

        From 1957 to 1958, he flew A3 aircraft while assigned
        to Heavy Attack Squadron Two deployed aboard the
        USS BON HOMME RICHARD and USS TICONDEROGA; and he
        was a research project pilot with Air Development
        Squadron Five until 1959. His assignment from
        1964 to 1965 was as Chief, Project Management
        Division of the Navy Field Office for Manned Orbiting
        Laboratory.

        He has accumulated 4,000 hours flight time --
        1,900 hours in jets.

        Captain Mitchell was in the group selected for
        astronaut training in April 1966. He served as
        a member of the astronaut support crew for Apollo 9
        and as backup lunar module pilot for Apollo 10.

        He completed his first space flight as lunar module
        pilot on Apollo 14, January 31 - February 9, 1971.
        With him on man's third lunar landing mission
        were Alan B. Shepard (spacecraft commander) and
        Stuart A. Roosa (command module pilot).

- more -

Manuevering their lunar module, "Antares," to a
landing in the hilly upland Fra Mauro region of
the Moon, Shepard and Mitchell subsequently deployed
and activated various scientific equipment and
experiments and collected almost 100 pounds of lunar
samples for return to Earth. Other Apollo 14 achievements include: first use of Mobile Equipment Transporter (MET); largest payload placed in lunar orbit;
longest distance traversed on the lunar surface;
largest payload returned from the lunar surface;
longest lunar surface stay time (33 hours); longest
lunar surface EVA (9 hours and 17 minutes); first
use of shortened lunar orbit rendezvous techniques;
first use of colored TV with new vidicon tube on
lunar surface; and first extensive orbital science
period conducted during CSM solo operations.

In completing his first space flight, Mitchell
logged a total of 216 hours and 42 minutes.

He was designated as backup lunar module pilot for
Apollo 16 on March 3, 1971.

- end -

March 1972

Spaceflight Tracking and Data Support Network

NASA's worldwide Spaceflight Tracking and Data Network
(STDN) will provide communication with the Apollo astronauts,
their launch vehicle and spacecraft. It will also maintain
the communications link between Earth and the Apollo experiments
left on the lunar surface and track the particles and fields
subsatellite injected into lunar orbit during Apollo 16.

The STDN is linked together by the NASA Communication
Network (NASCOM) which provides for all information and data
flow.

In support of Apollo 16, the STDN will employ 11 ground
tracking stations equipped with 9.1-meter (30-foot) and
25.9-meter (85-foot) antennas, an instrumented tracking ship,
and four instrumented aircraft. This portion of the STDN was
known formerly as the Manned Space Flight Network. For Apollo 16,
the network will be augmented by the 64-meter (210-foot) antenna
system at Goldstone, California (a unit of NASA's Deep Space
Network).

The STDN is maintained and operated by the NASA Goddard
Space Flight Center, Greenbelt, Md., under the direction of
NASA's Office of Tracking and Data Acquisition. Goddard will
become an emergency control center if the Houston Mission
Control Center is impaired for an extended time.

NASA Communications Network (NASCOM). The tracking network is linked together by the NASA Communications Network.
All information flows to and from MCC Houston and the Apollo
spacecraft over this communications system.

The NASCOM consists of more than 3.2 million circuit
kilometers (1.7 million nm), using satellites, submarine
cables, land lines, microwave systems, and high frequency
radio facilities. NASCOM control center is located at
Goddard. Regional communication switching centers are in
Madrid; Canberra, Australia; Honolulu; and Guam.

Intelsat communications satellites will be used for
Apollo 16. One satellite over the Atlantic will link Goddard
with Ascension Island and the Vanguard tracking ship. Another
Atlantic satellite will provide a direct link between Madrid
and Goddard for TV signals received from the spacecraft. One
satellite positioned over the mid-Pacific will link Carnarvon,

-more-

Canberra, Guam and Hawaii with Goddard through the Jamesburg, California ground station. An alternate route of communications between Spain and Australia is available through another Intelsat satellite positioned over the Indian Ocean if required.

Mission Operations: Prelaunch tests, liftoff, and Earth orbital flight of the Apollo 16 are supported by the Apollo subnet station at Merritt Island Fla., 6.4 km (3.5 nm) from the launch pad.

During the critical period of launch and insertion of the Apollo 16 into Earth orbit, the USNS Vanguard provides tracking, telemetry, and communications functions. This single sea-going station of the Apollo subnet will be stationed about 1610 km (870 nm) southeast of Bermuda.

When the Apollo 16 conducts the TLI maneuver to leave Earth orbit for the Moon, two Apollo range instrumentation aircraf (ARIA) will record telemetry data from Apollo and relay voice communications between the astronauts and the Mission Control Center at Houston. These aircraft will be airborne between Australia and Hawaii.

Approximately one hour after the spacecraft has been injected into a translunar trajectory, three prime MSFN stations will take over tracking and communication with Apollo. These stations are equipped with 25.9 meter (85-foot) antennas.

Each of the prime stations, located at Goldstone, Madrid, and Honeysuckle is equipped with dual systems for tracking the command module in lunar orbit and the lunar module in separate flight paths or at rest on the Moon.

For reentry, two ARIA (Apollo Range Instrumented Aircraft) will be deployed to the landing area to relay communications between Apollo and Mission Control at Houston. These aircraft also will provide position information on the Apollo after the blackout phase of reentry has passed.

An applications technology satellite (ATS) terminal has been placed aboard the recovery ship USS Ticonderoga to relay command control communications of the recovery forces, via NASA's ATS satellite. Communications will be relayed from the deck mounted terminal to the NASA tracking stations at Mojave, California and Rosman, N..C. through Goddard to the recovery control centers located in Hawaii and Houston.

-more-

Prior to recovery, the astronauts aeromedical records are transmitted via the ATS satellite to the recovery ship for comparison with the physical data obtained in the post-flight examination performed aboard the recovery ship.

Television Transmissions: Television from the Apollo spacecraft during the journey to and from the Moon and on the lunar surface will be received by the three prime stations, augmented by the 64-meter (210-foot) antennas at Goldstone and Parkes. The color TV signal must be converted at the MSC Houston. A black and white version of the color signal can be released locally from the stations in Spain and Australia.

Before the lunar surface TV camera is mounted on the LRV TV signals originating from the Moon will be transmitted to the MSFN stations via the lunar module. While the camera is mounted on the LRV, the TV signals will be transmitted directly to tracking stations as the astronauts explore the Moon.

Once the LRV has been parked near the lunar module, its batteries will have about 80 hours of operating life. This will allow ground controllers to position the camera for viewing the lunar module liftoff, post lift-off geology, and other scenes.

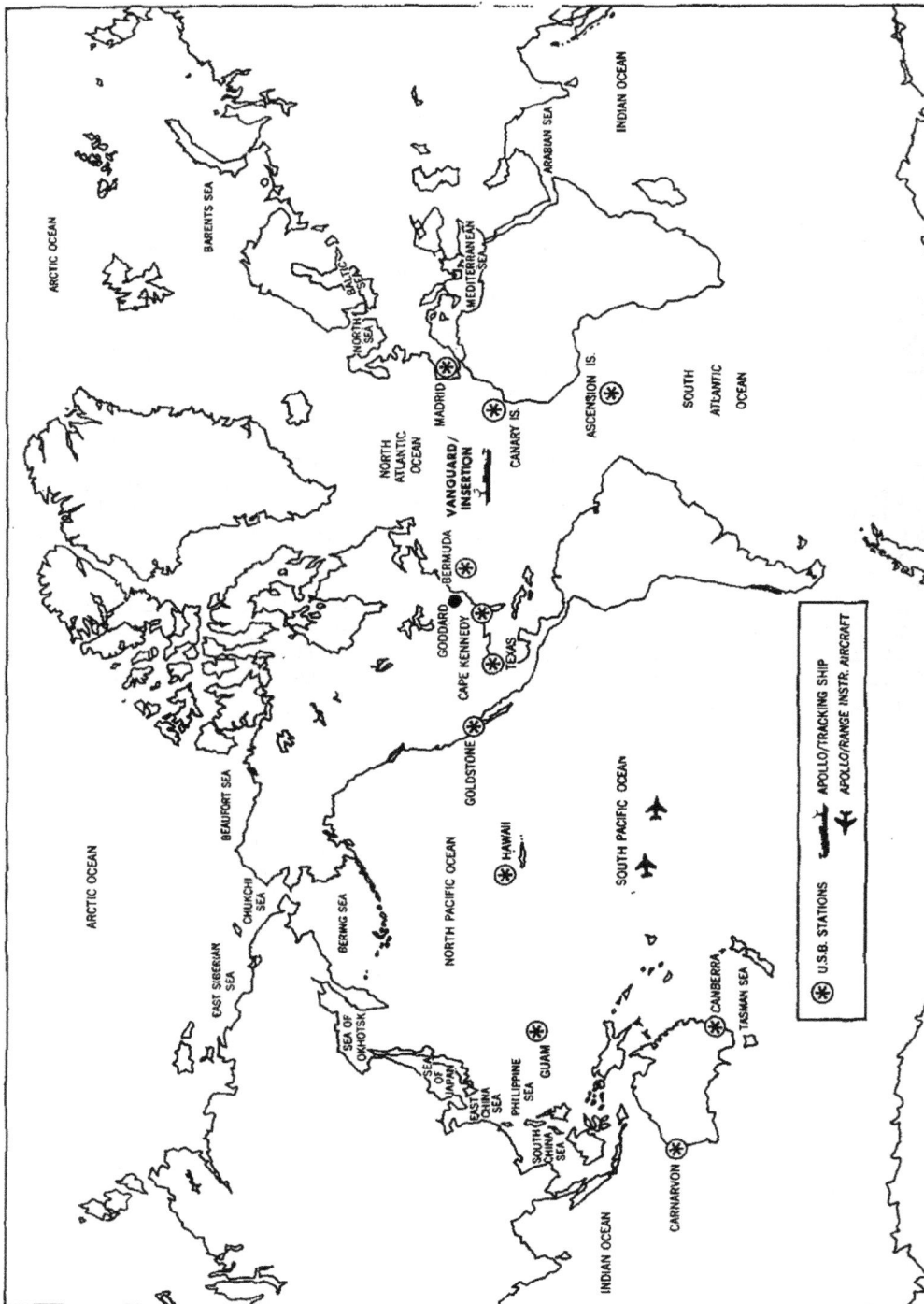

MANNED SPACE FLIGHT TRACKING NETWORK

ENVIRONMENTAL IMPACT OF APOLLO/SATURN V MISSION

Studies of NASA space mission operations have concluded that Apollo does not significantly effect the human environment in the areas of air, water, noise or nuclear radiation.

During the launch of the Apollo/Saturn V space vehicle, products exhausted from Saturn first stage engines in all cases are within an ample margin of safety. At lower altitudes, where toxicity is of concern, the carbon monoxide is oxidized to carbon dioxide upon exposure at its high temperature to the surrounding air. The quantities released are two or more orders of magnitude below the recognized levels for concern in regard to significant modification of the environment. The second and third stage main propulsion systems generate only water and a small amount of hydrogen. Solid propellant ullage and retro rocket products are released and rapidly dispersed in the upper atmosphere at altitudes above 70 kilometers (43.5 miles). This material will effectively never reach sea level and, consequently, poses no toxicity hazard.

Should an abort after launch be necessary, some RP-1 fuel (kerosene) could reach the ocean. However, toxicity of RP-1 is slight and impact on marine life and waterfowl are considered negligible due to its dispersive characteristics. Calculations of dumping an aborted S-IC stage into the ocean showed that spreading and evaporating of the fuel occurred in one to four hours.

There are only two times during a nominal Apollo mission when above normal overall sound pressure levels are encountered. These two times are during vehicle boost from the launch pad and the sonic boom experienced when the spacecraft enters the Earth's atmosphere. Sonic boom is not a significant nuisance since it occurs over the mid-Pacific Ocean.

NASA and the Department of Defense have made a comprehensive study of noise levels and other hazards to be encountered for launching vehicles of the Saturn V magnitude. For uncontrolled areas the overall sound pressure levels are well below those which cause damage or discomfort. Saturn launches have had no deleterious effects on wildlife which has actually increased in the NASA-protected areas of Merritt Island.

-more-

A source of potential radiation hazard but highly unlikely, is the fuel capsule of the radioisotope thermoelectric
generator supplied by the Atomic Energy Commission which
provides electric power for Apollo lunar surface experiments.
The fuel cask is designed to contain the nuclear fuel during
normal operations and in the event of aborts so that the
possibility of radiation contamination is negligible. Extensive
safety analyses and tests have been conducted which demonstrated
that the fuel would be safely contained under almost all credible accident conditions.

PROGRAM MANAGEMENT

The Apollo Program is the responsibility of the Office of Manned Space Flight (OMSF), National Aeronautics and Space Administration, Washington, D. C. Dale D. Myers is Associate Administrator for Manned Space Flight.

NASA Manned Spacecraft Center (MSC), Houston, is responsible for development of the Apollo spacecraft, flight crew training, and flight control. Dr. Christopher C. Kraft, Jr. is Center Director.

NASA Marshall Space Flight Center (MSFC), Huntsville, Ala., is responsible for development of the Saturn launch vehicles. Dr. Eberhard F. M. Rees is Center Director.

NASA John F. Kennedy Space Center (KSC), Fla., is responsible for Apollo/Saturn launch operations. Dr. Kurt H. Debus is Center Director.

The NASA Office of Tracking and Data Acquisition (OTDA) directs the program of tracking and data flow on Apollo. Gerald M. Truszynski is Associate Administrator for Tracking and Data Acquisition.

NASA Goddard Space Flight Center (GSFC), Greenbelt, Md., manages the Manned Space Flight Network and Communications Network. Dr. John F. Clark is Center Director.

The Department of Defense is supporting NASA during launch, tracking, and recovery operations. The Air Force Eastern Test Range is responsible for range activities during launch and down-range tracking. Recovery operations include the use of recovery ships and Navy and Air Force aircraft.

-more-

Apollo/Saturn Officials

NASA Headquarters

Dr. Rocco A. Petrone	Apollo Program Director, OMSF
Chester M. Lee (Capt., USN, Ret.)	Apollo Mission Director, OMSF
John K. Holcomb (Capt., USN, Ret.)	Director of Apollo Operations, OMSF
William T. O'Bryant (Capt., USN, Ret.)	Director of Apollo Lunar Exploration, OMSF

Kennedy Space Center

Miles J. Ross	Deputy Center Director
Walter J. Kapryan	Director of Launch Operations
Raymond L. Clark	Director of Technical Support
Robert C. Hock	Apollo/Skylab Program Manager
Dr. Robert H. Gray	Deputy Director, Launch Operations
Dr. Hans F. Gruene	Director, Launch Vehicle Operations
John J. Williams	Director, Spacecraft Operations
Paul C. Donnelly	Associate Director Launch Operations
Isom A. Rigell	Deputy Director for Engineering

Manned Spacecraft Center

Sigurd A. Sjoberg	Deputy Center Director, and Acting Director, Flight Operations
Brig. General James A. McDivitt (USAF)	Manager, Apollo Spacecraft Program
Donald K. Slayton	Director, Flight Crew Operations
Pete Frank	Flight Director
Phil Shaffer	Flight Director
Gerald D. Griffin	Flight Director
Eugene F. Kranz	Flight Director
Donald Puddy	Flight Director
Richard S. Johnston	Director, Medical Research and Operations (Acting)

Marshall Space Flight Center

Dr. Eberhard Rees	Director
Dr. William R. Lucas	Deputy Center Director, Technical
Richard W. Cook	Deputy Center Director, Management
James T. Shepherd	Director, Program Management

Herman F. Kurtz	Manager, Mission Operations Office
Richard G. Smith	Manager, Saturn Program Office
John C. Rains	Manager, S-IC Stage Project, Saturn Program Office
William F. LaHatte	Manager, S-II S-IVB Stages Project, Saturn Program Office
Frederich Duerr	Manager, Instrument Unit/GSE Project, Saturn Program Office
William D. Brown	Manager, Engine Program Office
James M. Sisson	Manager, LRV Project, Saturn Program Office

Goddard Space Flight Center

Ozro M. Covington	Director, Networks
William P. Varson	Chief, Network Computing & Analysis Division
H. William Wood	Chief, Network Operations Division
Robert Owen	Chief, Network Engineering Division
L. R. Stelter	Chief, NASA Communications Division

Department of Defense

Maj. Gen. David M. Jones, USAF	DOD Manager for Manned Space Flight Support Operations
Col. Alan R. Vette, USAF	Deputy DOD Manager for Manned Space Flight Support Operations, and Director, DOD Manned Space Flight Support Office
Rear Adm. Henry S. Morgan, Jr., USN	Commander, Task Force 130, Pacific Recovery Area
Rear Adm. Roy G. Anderson, USN	Commander Task Force 140, Atlantic Recovery Area
Capt. E. A. Boyd, USN	Commanding Officer, USS Ticonderoga, CVS-14 Primary Recovery Ship
Brig. Gen. Frank K. Everest, Jr., USAF	Commander Aerospace Rescue and Recovery Service

-more-

Herman F. Kurtz	Manager, Mission Operations Office
Richard G. Smith	Manager, Saturn Program Office
John C. Rains	Manager, S-IC Stage Project, Saturn Program Office
William F. LaHatte	Manager, S-II S-IVB Stages Project, Saturn Program Office
Frederich Duerr	Manager, Instrument Unit/GSE Project, Saturn Program Office
William D. Brown	Manager, Engine Program Office
James M. Sisson	Manager, LRV Project, Saturn Program Office

Goddard Space Flight Center

Ozro M. Covington	Director, Networks
William P. Varson	Chief, Network Computing & Analysis Division
H. William Wood	Chief, Network Operations Division
Robert Owen	Chief, Network Engineering Division
L. R. Stelter	Chief, NASA Communications Division

Department of Defense

Maj. Gen. David M. Jones, USAF	DOD Manager for Manned Space Flight Support Operations
Col. Alan R. Vette, USAF	Deputy DOD Manager for Manned Space Flight Support Operations, and Director, DOD Manned Space Flight Support Office
Rear Adm. Henry S. Morgan, Jr., USN	Commander, Task Force 130, Pacific Recovery Area
Rear Adm. Roy G. Anderson, USN	Commander Task Force 140, Atlantic Recovery Area
Capt. E. A. Boyd, USN	Commanding Officer, USS Ticonderoga, CVS-14 Primary Recovery Ship
Brig. Gen. Frank K. Everest, Jr., USAF	Commander Aerospace Rescue and Recovery Service

CONVERSION TABLE

	Multiply	By	To Obtain
Distance:	inches	2.54	centimeters
	feet	0.3048	meters
	meters	3.281	feet
	kilometers	3281	feet
	kilometers	0.6214	statute miles
	statute miles	1.609	kilometers
	nautical miles	1.852	kilometers
	nautical miles	1.1508	statute miles
	statute miles	0.8689	nautical miles
	statute miles	1760	yards
Velocity:	feet/sec	0.3048	meters/sec
	meters/sec	3.281	feet/sec
	meters/sec	2.237	statute mph
	feet/sec	0.6818	statute miles/hr
	feet/sec	0.5925	nautical miles/hr
	statute miles/hr	1.609	km/hr
	nautical miles/hr (knots)	1.852	km/hr
	km/hr	0.6214	statute miles/hr
Liquid measure, weight:	gallons	3.785	liters
	liters	0.2642	gallons
	pounds	0.4536	kilograms
	kilograms	2.205	pounds
	metric ton	1000	kilograms
	short ton	907.2	kilograms
Volume:	cubic feet	0.02832	cubic meters
Pressure:	pounds/sq. inch	70.31	grams/sq.cm
Thrust:	pounds	4.448	newtons
	newtons	0.225	pounds
Temperature:	Centigrade	1.8; add 32	Fahrenheit

www.ingramcontent.com/pod-product-compliance
Lightning Source LLC
Chambersburg PA
CBHW051215200326
41519CB00025B/7123

* 9 7 8 1 7 8 0 3 9 8 6 5 5 *